WHY GEOGRAPHY MATTERS

WHY GEOGRAPHY MATTERS
MORE THAN EVER

Harm de Blij

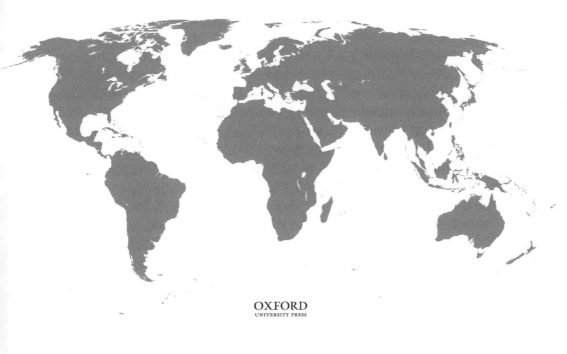

OXFORD
UNIVERSITY PRESS

OXFORD
UNIVERSITY PRESS

Oxford University Press is a department of the University of Oxford. It furthers the University's
objective of excellence in research, scholarship, and education by publishing worldwide.

Oxford   New York
Auckland   Cape Town   Dar es Salaam   Hong Kong   Karachi
Kuala Lumpur   Madrid   Melbourne   Mexico City   Nairobi
New Delhi   Shanghai   Taipei   Toronto

With offices in
Argentina   Austria   Brazil   Chile   Czech Republic   France   Greece
Guatemala   Hungary   Italy   Japan   Poland   Portugal   Singapore
South Korea   Switzerland   Thailand   Turkey   Ukraine   Vietnam

Oxford is a registered trademark of Oxford University Press in the UK and certain other countries.

Published in the United States of America by
Oxford University Press
198 Madison Avenue, New York, NY 10016

Library of Congress Cataloging-in-Publication Data
De Blij, Harm J.
Why geography matters : more than ever / Harm de Blij. — 2nd ed.
p.   cm.
ISBN 978-0-19-991374-9
1. Human geography—United States—History—21st century. 2. Terrorism—History—21st century.
3. Climatic changes—History—21st century. 4. China—Politics and government—2002– 5. United
States—Politics and government—2001–2009. 6. United States—Social conditions—21st century. I. Title.
GF503.D4 2012
909.83—dc23   2012011011

ISBN 978-0-19-991374-9

3   5   7   9   8   6   4   2

Printed in the United States of America on acid-free paper.

DEDICATION

I have been enormously fortunate to have benefited from the talents, knowledge and insights of numerous colleagues, twenty-one of whom joined me to co-author books and articles during my 55-year career in geography. I gratefully and affectionately dedicate *More Than Ever* to them, with thanks beyond measure:

Alan Best	Boston University*
Steve Birdsall	University of North Carolina
Jim Burt	University of Wisconsin
Don Capone	University of Miami
Roy Cole	Grand Valley State University
Gerald Danzer	University of Illinois, Chicago Circle
Dorothy Drummond	Indiana State University
Erin Fouberg	Northern State University
Marty Glassner	Southern Connecticut University
David Greenland	University of Colorado
Roger Hart	City University of New York
John Hunter	Michigan State University
Esmond Martin	Independent Scholar, Nairobi
Joe Mason	University of Wisconsin
Peter Muller	University of Miami
Alec Murphy	University of Oregon
Jan Nijman	University of Amsterdam
Gary Peters	California State University, Long Beach
Eric Spears	Mercer University
Richie Williams	Woods Hole Research Institute
Antoinette WinklerPrins	Michigan State University

*Geographers tend to be mobile: institutional affiliations refer to the time of collaboration.

PREFACE

To answer the question posed by the title of this book: geographic illiteracy poses a huge risk to America's national security. As to the subtitle, policy options available in pre-recession years have disappeared in a time of shrinking budgets and growing challenges. Geographic knowledge matters more than ever.

There is ample evidence that inadequate geographic comprehension contributes significantly to flawed policy making. Robert McNamara's anguished retrospective on the Vietnam War, citing misconceptions about the physical and human geography of Indochina in the planning and execution of that costly campaign, constitutes prima facie evidence (McNamara, 1995). Furthermore, McNamara argued, military and civilian leaders did a poor job of informing the American public of the social "realities" in Vietnam and the physical and political stage on which the contest played itself out. It hasn't gotten any better. Apart from the nonsense about Iraq's purported weapons of mass destruction, the implications of American intervention in a sectarian-fractured Islamic country where brigades of armed and uniformed infidels would be greeted with flowers and gifts evinced a disconnect with reality that misled the country and doomed the effort. As to informing the public, several years into the war a national newspaper survey revealed that only one in seven Americans could identify Iraq on a blank map of Eurasia. In 2010, the Chair of an important Senate committee, returning from Kabul, noted on television that Afghanistan was not divided ethnically, "like you have in Iraq" (see page 184). Tell that to the Shi'ite Hazaras or the always-suspect Tajiks in a Puhstun-dominated, Taliban-infested, Sunni-majority state.

It is conceivable that the invasion of smaller countries by larger powers (such as Russia in Georgia: the United States was not alone) will become a less-common phenomenon under more austere fiscal circumstances, but a greater threat looms: a new bipolar world in which two Pacific powers with discordant objectives and capacities may find themselves in a contest that has the potential to precipitate a new Cold War. Chinese leaders have a clear vision of the new political geography they will seek: it requires a withdrawal of American presence and influence from waters and islands of the Western Pacific. Americans' perceptions of that political geography are weaker: it is farther away, historically stable, not yet clear enough to rouse passions. This asymmetry of geographic awareness and knowledge may be the most perilous of all.

These developments play themselves out on an environmental stage that is in a transition of its own. An earlier version of this largely-rewritten book cited climate change as a crucial factor capable of precipitating global instability, but there can be no doubt that its urgency has declined among priorities that dominate public opinion. It nevertheless remains a key factor in my argument as climate change may alter economic and political prospects in critical ways.

The dominant message of *Why Geography Matters*, though, involves responsibility: as the democratic nation that elects governmental representatives whose decisions affect not just America but the entire world, Americans have a particular obligation to be well informed about our small and functionally shrinking planet. But surveys show that this obligation is not being met, a circumstance that is underscored by steadily declining media coverage of international news. Between the early 1980s and the start of the current decade and measured in terms of minutes devoted to it, such coverage declined by three-quarters. During the interminable presidential-primary campaign of 2011–2012, international news faded from view for days at a time.

Looking at the world geographically—that is, comprehensively—tends to lead to what my colleagues at ABC Television's Good Morning America used to call the "wow factor": unexpected explanations or insights into seemingly intractable problems, making geography a great way to understand our complex world. Geography has a way of coming up with unforeseen linkages, between climate change and historical events, between natural phenomena and political developments, between environment and behavior, that are unmatched in other fields. And geography tends to look at the here and now, and at the future. Coupled with Americans' still-low level of geographic literacy is a kind of cultural obsession with the past—especially the American past—that crowds out other perspectives. It is inconceivable to go through school or college without required courses in history, but few of us graduate with some exposure to modern geography. Far too few.

If I may make a suggestion: when reading this book, please have a good world atlas handy. This book contains maps, but you can never have too many. And a globe would be useful as well. No home should be without one. Should you find yourself wanting to delve more deeply into geography, a textbook listed in the citations contains more than 200 thematic color maps drawn by Mapping Specialists, whose excellent work has supported my writings for many years (de Blij, Muller, and Nijman, 2012).

I had the good fortune to encounter geography as a future profession early in life, and that good fortune was compounded by the colleagues I met and the friends I made among the small but vibrant community of geographers working in a range of professional settings that would astound you (from the wine industry to NOAA, from the travel industry to teaching, from the publishing industry to government). Many of these colleagues and friends, some in related academic disciplines, contributed in one way or another to the first version of this book, or did so for this iteration, by bringing developments and ideas to my attention, by debating content, by reading drafts, sometimes by making a brief comment that led to unanticipated insights. Some of this assistance is reflected by the citations, but here I express my appreciation to Dick Allen, Barbara Bailey, Richard Banz, Bob Begam, Phil Benton, Alan Best, Dick Boehm, Dave Campbell, Iraphne Childs, Pinkie Christensen, Spencer Christian, Roy Cole, Bob Crook, Tanya de Blij, George Demko, Ryan Flahive, Erin Fouberg, Gary Fuller, Charlie Gibson, Ed Grode, Dick Groop, Gil Grosvenor, Jay Harman, Carl Hirsch, Chip Hoehler, John Hunter, Tom Jeffs, Arte Johnson, Grady Kelly-Post, Jim King, Peter Krogh, John Larsen, Neal Lineback, Don Lubbers, John Malinowski, Geof Martin, Patricia McCulley, Dave McFarland, Paul McFarland, Betty Meggers, Hugh Moulton, Peter Muller, Alec Murphy, David Murray, Jim Newman, Dan Niemeyer, Jan Nijman, Eugene Palka, Charlie Pirtle, Mac Pohn, Jim Potchen, Frank Potter, Victor Prescott, Doug Richardson, Jack Reilly, Dick Robb, Chris Rogers, Eunice Rutledge, Stathis Sakellarios, Birthe Sauer, Jeff Schaffer, Ruth Sivard, Thomas Sowell, Peter Spiers, Mary Swingle, Derrick Thom, Morris Thomas, George Valanos, Norman Valiant, Jack Weatherford, Frank Whitmore, Antoinette WinklerPrins, Duke Winters, and Henry Wright. I acknowledge appreciatively the critical review of this narrative by Richard Earl of Texas State University and greatly appreciate noted cartoonist Ed Hammond's permission to include on page 144 the cartoon he published following one of my lectures at The Chautauqua Institution. I am, of course, solely responsible for the contents of this book, and nothing in it implies endorsement by all or any of those just cited.

Once again I am grateful to Don Larson and his team at Mapping Specialists in Madison, Wisconsin for their splendid work as they made the most of limited cartographic options.

My former editor at Oxford University Press, Ben Keene, proposed that I write this book and combined his intense interest in it with his creative energies. I am fortunate to have once again benefited from his editorial skills. I further acknowledge appreciatively the work of Production Editor Natalie Johnson, Executive Editor Timothy Bent, and his editorial assistant, Keely Latcham.

Harm de Blij

CONTENTS

WHY GEOGRAPHY MATTERS

1

WHY GEOGRAPHY MATTERS . . . MORE THAN EVER!

KNOWLEDGE, AS THE FATES OF HUMAN SOCIETIES HAVE DEMON-strated countless times, is power. Whether such knowledge involves an understanding of the seasonal cycles of natural irrigation or the capacity to find and exploit hidden energy reserves, or entails inventions ranging from agricultural tools to sophisticated weapons, it has spelled advantage in an ever-more competitive world.

Not only is our world ever more competitive: it is also changing at an ever-faster pace. National as well as local governments must make decisions in short order as global and regional challenges arise at breakneck speed. In mid-December 2010, an incident in a market in Tunisia catapulted a seemingly stable member of the Arab League into a full-scale revolution that toppled its government in less than one month, starting a sequence of events in other countries that soon became known as the "Arab Spring." Six months later, this Internet-propelled movement had spread from Morocco in the west to Bahrain in the east, and a civil war was in progress in Libya and looming in Syria. Non-Arab states suddenly found themselves having to take sides; in the case of Libya, the issues ranged from the level of support for anti-government rebels (weapons? money?) to military involvement (ground troops? bombings?). Such decisions must be based on available knowledge of conditions not only in Libya but also in other countries affected by the "Arab Spring," and these conditions comprise a host of circumstances: cultural, political, economic, environmental. Certainly there are specialists and experts in these outsider countries who can advise members of the American and

European (and other) administrations on local circumstances, but in the end the decisions are made by elected representatives in government. And then the question becomes: how well informed are *they*?

The answer is not encouraging. Listen to the commentaries by members of the United States Congress on those Sunday-morning television talk shows, and you often cringe at what you hear. True, our representatives have to deal with many and diverse issues, but it's obvious that, when it comes to the wider world, their knowledge is often fragmentary.

Given the accelerating pace of change on our increasingly crowded planet, this may not be surprising—even if it is disturbing. Just consider crucial events in the first decade-plus of the twenty-first century: intense climate change accompanied by significant weather extremes; deadly tsunamis caused by submarine earthquakes; unprecedented terrorist attacks in the United States, Europe and elsewhere; costly wars in Iraq and Afghanistan; a terrible, mostly overlooked conflict in Equatorial Africa costing millions—yes, millions—of lives; an economic crisis threatening the stability of the international system even as it throws the United States into recession. Add to this the burgeoning presence of China on the international stage and the growing role of India, the specter of troubling disarray in the European Union, and concern over nuclear ambitions in North Korea and Iran, and it is obvious that the wider world presents daunting challenges for decision makers.

All this is happening right after one of the most tumultuous periods in world history, witnessing the collapse of the Soviet Union and Yugoslavia and their combined disintegration into some two dozen new states, the momentous transition to democracy in South Africa, the emergence of NAFTA, and the waging of what became known as the Gulf War. Time and again, during those last two decades of the twentieth century, the map of the world changed drastically, to the point that the makers of expensive and bulky globes mostly gave up. And it isn't over, although in terms of state disintegration, the pace of change has at least slowed down. South Sudan in 2011 became the newest officially recognized state on the map, the 193rd member of the United Nations, and in many ways the poorest. All members of the international community acknowledged South Sudan's independence, unlike Kosovo (in Europe, the latest fragment of the former Yugoslavia to seek sovereignty, is recognized by many but not all).

Is there a conceptual framework that can accommodate all these changes, that would help us understand the transformations and interconnections, inform our thoughts and decisions through a particular, comprehensive perspective? This book answers these questions with one affirmation: geography.

In truth, geography itself has gone through several transformations in recent times. When I was a high-school student, learning to name countries and cities, ranges and rivers, was an end in itself. Making the connections that give geography its special place among the sciences was not on the agenda. By the time I got to college, geography (in Europe and America at least) had become more scientific, even mathematical. During my teaching career it became more technological, and not for nothing does the now-common acronym GIS stand for Geographic Information Systems. Today geography has numerous dimensions, but it remains a great way to comprehend our complex world.

BECOMING A GEOGRAPHER

Not long ago I read an interview with a prominent geographer in the newsletter of this country's largest professional geographic organization. The editor asked Frederick E. (Fritz) Nelson, now teaching at the University of Delaware, a question all of us geographers hear often: what caused you to join our ranks? His answer is one given by many a colleague: while an undergraduate at Northern Michigan University he took a course in regional geography and liked it so much that he decided to pursue a major in the discipline. He changed directions while a graduate student at Michigan State University, but he did not forget what attracted him to geography originally. Today his research on the geography of periglacial (ice-margin) phenomena is world renowned (Solis, 2004).

My own encounter with geography stems from my very first experience with it in Holland during the Second World War, not at school, but at home. With my dad I watched in horror from a roof window in our suburban house when my city, Rotterdam, was engulfed by flames following the nazi fire-bombing of May 14, 1940 (long-buried feelings that resurfaced on September 11, 2001), and soon my parents abandoned the city for a small village near the center of the country. There they engaged in a daily battle for survival, and I spent much time in their library, which included several world and national atlases, a large globe, and the books of a geographer named Hendrik Willem van Loon. As the winters grew colder and our situation deteriorated, those books gave me hope. Van Loon described worlds far away, where it was warm, where skies were blue and palm trees swayed in soft breezes, and where food could be plucked from trees. There were exciting descriptions of active volcanoes and of tropical storms, of maritime journeys to remote islands, of great, bustling cities, of powerful kingdoms and unfamiliar customs. I traced van Loon's journeys on atlas maps and dreamed of the day when I would see his worlds for myself. Van Loon's geographies gave me, almost literally, a lease on life.

After the end of the war, my fortunes changed in more than one way. When the schools opened again, my geography teacher was an inspiring taskmaster who made sure that we, a classroom full of youngsters with a wartime gap in our early education, learned that while geography could widen our horizons, it also required some rigorous studying. The rewards, he rightly predicted, were immeasurable.

If, therefore, I write of geography with enthusiasm and in the belief that it can make life easier and more meaningful in this complex and changing world, it is because of a lifetime of discovery and fascination.

WHAT IS GEOGRAPHY?

As a geographer, I've often envied my colleagues in such fields as history, geology, and biology. It must be wonderful to work in a discipline so well defined by its name and so accurately perceived by the general public. Actually, the public's perception may not be so accurate, but people *think* they know what historians, geologists, and biologists do.

We geographers are used to it. Sit down next to someone in an airplane or in a waiting room somewhere, get involved in a conversation, and that someone is bound to ask: Geography? You're a geographer? What is geography, anyway?

In truth, we geographers don't have a single, snappy answer. A couple of millennia ago, geography essentially was about discovery. A Greek philosopher named Eratosthenes moved geographic knowledge forward by leaps and bounds; by measuring Sun angles, he not only concluded that the Earth was round but came amazingly close to the correct figure for its circumference. Several centuries later, geography was propelled by exploration and cartography, a period that came to a close, more or less, with the adventures and monumental writings of Alexander von Humboldt, the German naturalist-geographer. A few decades ago, geography still was an organizing, descriptive discipline whose students were expected to know far more capes and bays than were really necessary. Today geography is in a new technological age, with satellites transmitting information to computers whose maps are used for analysis and decision making.

Despite these new developments, however, geography does have some traditions. The first, and in many ways the most important, is that geography deals with the natural as well as the human world. It is, therefore, not just a "social" science. Geographers do research on glaciations and coastlines, on desert dunes and limestone caves, on weather and climate, even on plants and animals. We also study human activities, from city planning to boundary making, from winegrowing to churchgoing. To me, that's the best part of geography: there's almost nothing in this wide, wonderful world of ours that can't be studied geographically.

This means, of course, that geographers are especially well placed to assess the complicated relationships between human societies and natural environments; this is geography's second tradition. In this arena knowledge is fast growing, and if you want to see evidence of the insights geography can contribute I know of no better book than Jean Grove's spellbinding analysis of what happened when Europe and much of the rest of the world were plunged into what she calls *The Little Ice Age*, starting around 1300 and continuing, with a few letups, until the early 1800s (Grove, 2004). This is a global, sweeping analysis; other geographers work at different levels of scale. Some of my colleagues study and predict people's reactions to environmental hazards: Why do people persist in living on the slopes of active volcanoes and in the floodplains of flood-prone rivers? How much do home buyers in California know about the earthquake risk at the location of their purchase and what are they told by real-estate agents before they buy? Another environment-related issue involves health and disease. The origins and spread of many diseases have much to do with climate, vegetation, and fauna as well as cultural traditions and habits. A small but productive cadre of medical geographers is at work researching and predicting outbreaks and dispersals of maladies ranging from cholera to AIDS to bird flu. Peter Gould's book on AIDS, which he called *The Slow Plague*, effectively displays the toolbox of geographers when it comes to such analyses (Gould, 1993).

A third geographic tradition is simply this: we do research in, and try to understand, foreign cultures and distant regions. A few decades ago, it was rare to find a geographer who did not have some considerable expertise in a foreign area, large or small. Most spoke one or more foreign languages (this used to be a requirement for graduation with a doctorate), kept up with the scholarly literature as well as the popular press in their chosen region, and conducted repeated research there. That tradition has faded somewhat in the new age of the Internet, satellite data, and computer cartography, but many students still are first attracted to geography because it aroused their curiosity about some foreign place. The decline in interest in international affairs is not unique to geography, of course. From analyses of network news content to studies of foreign-area specialization in United States intelligence operations, our isolationism and parochialism are evident. But there will be a rebound, probably of necessity more than desire. Geographic provincialism entails serious national security risks.

A fourth tradition geographers like to identify is the so-called location tradition, which is essentially a human-geographic (not a physical-geographic) convention. Why are activities, such as movie industries or shopping centers, or towns or cities such as Sarasota, Florida, or Tokyo, Japan located where they are? What does their location imply for their prospects? Why did one city thrive and grow while a nearby settlement dwindled and failed? Often a geographic answer illuminates historic events. Urban and regional planning is now a key component to many college geography curricula, and many of our graduates find positions in the planning field.

It is true that these traditional geographies have helped unite as well as divide geography and geographers. If they form a unifying element, geography's broad umbrella can also lead to separation. It's a long stretch from glacial landforms to urban structure, from soil distribution to economic models, and specialization has a way of eroding the common ground.

But take heart: the technological revolution propelled by the Internet has ushered in a new era in geographic research and analysis. In Chapter 2, we look at the changing role of maps and cartography in geographic education, investigation, interpretation, and demonstration, and the acronym GIS—Geographic Information Systems—the technology that has not only revolutionized geographic inquiry but has coalesced the discipline as never before. Correlations involving distributions of apparently disparate phenomena, that in pre-GIS times would have taken months to achieve, can now be accomplished in minutes. The urban specialist who might never have thought of glacial landforms beneath city streets can now see them for herself, the latest information the click of a mouse away. In the process, we learn what our colleagues in other fields are doing, thus gaining knowledge about new directions in geography, wherever these may be leading.

LOOKING AT THE WORLD SPATIALLY

If there is one word that telegraphs the thinking that underpins geography's traditions, methods, and technologies, that word would derive from space—not celestial space, but Earthly space. We geographers look at the world *spatially*. I sometimes try this concept on a questioner: historians look at the world temporally or chronologically; economists and political scientists come at it structurally, but we geographers look at it spatially. With a little luck my interrogator will furrow a brow, nod understandingly, and take out his or her *USA Today* and read about the results of the latest geographic literacy test.

Geographers, of course, are not the only scholars to use spatial analysis to explain the workings of our world. Economists, anthropologists, and other social scientists sometimes take a spatial perspective as well although, as their writings suggest, they often lag behind. Geographers were amused (a few were annoyed) when the noted economist Paul Krugman began writing his columns in the *New York Times* and rediscovered spatial truisms that had long since been superseded in the geographic literature (Berry, 1999). The physiologist Jared Diamond's magisterial book *Guns, Germs, and Steel* was described by *New York Times* journalist John N. Wilford as "the best book on geography in recent years," but geographers noted some significant conceptual weaknesses in it (Diamond, 1997). Mr. Diamond not only took note of these caveats, but acted impressively on them: he joined the faculty of the Department of Geography at UCLA and wrote a successor volume that demonstrates his perception of geographic factors in the disintegration of once-thriving societies (Diamond, 2005).

Diamond, in both of these Herculean works, raised sensitive issues that once lay at the core of geographic research: the role of natural environments in the fate of human societies. Early in the twentieth century, this research led to generalizations attributing the "energy" of midlatitude societies and the "lethargy" of tropical peoples to climate. Such simplistic analyses were not only bound to be flawed, but could be used to give credence to racist interpretations of the state of the world, discrediting the whole enterprise. But the fundamental question, as Diamond asserted, has not gone away. Today we know a great deal more about environmental swings and associated ecological transitions as well as human dispersal and behavior, and the issue is getting renewed attention.

Nevertheless, it remains tempting to assign a simple causal relationship to a complex set of circumstances because a map suggests it. Consider the following quote from a lecture presented at the United States Naval War College by another noted economist, Jeffrey Sachs: "Virtually all of the rich countries of the world are outside the tropics, and virtually all of the poor countries are within them . . . climate, then, accounts for quite a significant proportion of the cross-national and cross-regional disparities of world income" (Sachs, 2000). That would seem to be a reasonable conclusion, but the condition of many of the world's poor countries results from a far more complex combination of circumstances including subjugation, colonialism, exploitation, and suppression that put them at a disadvantage that will long endure and for which climate may not be the significant causative factor Mr. Sachs implies. In any case, while it is true that many of the world's poor countries lie in tropical environs, many others, from Albania to Turkmenistan and from Moldova to North Korea, do not. The geographic message does not lend itself to environmental generalizations.

Of course we should be pleased that nongeographers are jumping on our bandwagon, but this does not make our effort to come up with a generally accepted definition of our discipline any easier. In some ways, I suppose, this very difficulty is one of geography's strengths. Geography is a discipline of diversity, under whose "spatial" umbrella we study and analyze processes, systems, behaviors, and countless other phenomena that have spatial expression. It is this tie that binds geographers, this interest in patterns, distributions, diffusions, circulations, interactions, juxtapositions—the ways in which the physical and human worlds are laid out, interconnect, and interact. Yes, it is true that some tropical environs are tough on farmers and engender diseases. Tougher still, though, are the rich world's tariff barriers against the produce of tropical-country farmers and the subsidies paid to large agribusiness. End those practices, and suddenly climate won't seem so "significant" a factor in the global distribution of poverty.

So geography's umbrella is large, allowing geographers to pursue widely varying research. These days that includes a lot of social activism and other work that might seem closer to sociology than to geography, but much geographic research remains spatial and substantive. I have colleagues whose work focuses on Amazonian deforestation, West African desertification, Asian economic integration, Indonesian transmigration. Others take a more specific look at such American phenomena as professional football and the sources and team destinations of players, the changing patterns of church membership and evangelism, the rise of the wine industry in this period of global warming, and the impact of NAFTA on manufacturing employment in the Midwest. I'm always fascinated to read in our professional journals what they're discovering, and as I used to tell my students, the Age of Discovery may be over, but the era of geographic discovery never will be.

THE SPATIAL SPECIALIZERS

The stirring story of geography's early emergence, its Greek and Roman expansion, its European diversification, and its global dissemination is a saga of pioneering observation, heroic exploration, inventive mapmaking, and ever-improving interpretation, discussed in fascinating detail by the discipline's leading historian (Martin, 2005). Long before European colonialism launched the first wave of what today we call globalization, indigenous geographers were drawing maps and interpreting landscapes from Korea to the Andes and from India to Morocco. Later, geographic philosophy got caught up in European nationalism, and various "schools" of geography—German, French, British—came to reflect, and even to support and justify, national political and strategic aspirations including expansionism, colonialism, and even naziism. In the United States, geography also generated specialized schools of thought, but the issues that defined (and divided) them tended to be scholarly rather than political. The most prominent of these American schools was based for many years at the University of California–Berkeley, and was dominated by the powerful personality of the cultural geographer Carl Sauer. The core idea of this school was the notion that a society's lifeways would be imprinted on the Earth as a *cultural landscape* that could be subjected to spatial analysis wherever it was found.

Geographers not only take a wide view, but also a long view. We try not to lose sight of the forest for the trees, and put what we discover in temporal as well as spatial perspective. "Geography is synthesis," is one fairly effective answer to that question about just what geography is. That is, geographers try to find ways to link apparently disparate information to solve unanswered problems. As you will see later, sometimes such daring generalizations can set research off in very fruitful directions.

These days, though, it takes courage to generalize and hypothesize. This, as we all know, is the age of specialization. But specialized research ought to have some link to the big questions that confront us, or you have reason to question its value. Fifty years ago one of my professors at Northwestern University often urged me and my co-students to practice what he called "intelligent dinner conversation" (a quaint cultural tradition, remnants of which are still observable in certain urban settings). "Always be ready to explain in ordinary language to the guest across the table what it is you do and why it matters," he said. Most of us thought that this was not only unnecessary, but also none of "the public's" business. But he was right, and he would enjoy the debate now going on in professional geography, much of which focuses on ways to speak to the general public in plain language about what it is we do.

Specialization in research and teaching occurs at several levels, of course. I have already mentioned that some geographers (fewer than before) still become area specialists or, in another context, regional scientists. Others study urbanization from various spatial standpoints, and their studies range from highly analytical research on land values and rents to speculative assessments of intercity competition. One especially interesting question has to do with efforts to measure the amount of interaction between cities. When two large cities lie fairly close together, say Baltimore and Philadelphia, there will be more interaction (in numerous spheres ranging from telephone calls to road traffic) than when two cities lie much farther apart, for example Denver and Minneapolis. But just how does this level of interaction vary with city size and intercity distance? The answer is embodied by the so-called gravity model, which holds that interaction can be represented by a simple formula: multiply the two urban populations and divide the total by the square of the distance between them. You can use kilometers, miles, or even some other measure of distance, but so long as you are consistent for comparative purposes the model will do a good job of predicting reality. Distance is a powerful deterrent to interaction—geographers call this distance decay—and measuring this factor can be enormously helpful in business and commercial decision making.

Other geographers combine economics and geography, and focus on spatial aspects of economic activities. The rise of the world's new economic giants on the Pacific Rim has kept them busy.

Still others focus on spatial aspects of political behavior. Political scientists tend to focus on institutions, political geographers on political mosaics. Geopolitics, an early subfield of political geography, was hijacked by nazi ideologues and lost its reputation; but recently, geopolitics has been making a comeback as an arena of serious and objective research. From power relationships to boundary studies, political geography is a fascinating field.

There are literally dozens of fields of specialization in geography, and for students contemplating a career in geography it's a little bit like being in a candy store. Interested in anthropology? Try cultural geography! Biology? There's biogeography! Geology? Don't forget geomorphology, the study of the evolution of landscape. Historical geography is an obviously fruitful alliance between related disciplines. The list of such options is long, and it is still growing. Developments in mapmaking have opened whole new horizons for technically inclined geographers.

Over a lifetime of geographic endeavor, many geographers change specialties, and I'm one of them. I was educated to be a physical geographer, that is, as someone who specializes in landscape study (geomorphology) and related fields. As such, I spent a year in the field in Swaziland, in southern Africa, trying to determine whether a large, wide valley there was a part of the great African Rift Valley system, the likely geographic source of humanity. While I was preparing for this research, however, I met a political geographer named Arthur Moodie, a British scholar who came to Northwestern University as a visiting professor. I took his classes and never forgot them. When I was hired by Michigan State University as a physical geographer, I also continued to read and study political geography. Eventually, I was asked to teach a course in that field, wrote a book and some articles about it, and thus developed a second specialization.

What I didn't realize, at first, was how my background in physical geography would make me a better political geographer. Like geopolitics, environmental determinism had acquired a bad name between the World Wars, and it could be dangerous, professionally, to try to explain political or other social developments as influenced by environmental circumstances. But I knew that, like geopolitics, environmental studies would make a comeback. When they did, I had the background to participate in the debate. That's how, many years later, I was appointed to Georgetown University to teach environmental issues in the School of Foreign Service.

I made only one other foray into a new field, and that was also as pleasant a geographic experience as I've ever had. It all began with a great bottle of wine. A fateful dinner with that bottle of 1955 Chateau Beychevelle so aroused my curiosity that, five years later, I was working on my book entitled *Wine: A Geographic Appreciation*, was teaching a course called The Geography of Wine at the University of Miami, and saw some of my students enter the wine business armed with a background they often found to be very advantageous. Geography has few limits—and specialization does have its merits.

BUT IS GEOGRAPHY IMPORTANT?

Remember the bumper sticker, popular some years ago, that said "If You Can Read This, Thank a Teacher"? One day I was driving down one of my least favorite highways, Interstate 95 between Fort Lauderdale and Miami in Florida, when a car passed me whose owner had modified that sticker by inserting the word "Map" after "This" and by pasting a piece of road map at the end of the slogan. I didn't need to ask what that owner's profession was. A geography teacher, obviously.

The fact is, a lot of us cannot read maps. Surveys show that huge numbers of otherwise educated people don't know how to use a map effectively. Even simple road maps are beyond many more of us than you might imagine. People who, you would think, deal with maps all the time and therefore know how to get the most out of them—travel professionals—often have trouble with maps. I live about half the year on Cape Cod, and thus have the dubious pleasure of flying into and out of Boston's Logan Airport, about two hours from home. These days flight schedules are not what they used to be, so when someone arranges my trip I always hope that consideration was given to the other airport about two hours from the mid-Cape, Providence. I've learned not to count on it.

Anomalously, the now-widespread availability of GPS (Global Positioning Systems) equipment, hand-held and built into automobile dashboards, seems to be having an unexpectedly negative effect on orientation and awareness. A recent, and as-yet-anecdotal, press report commented on the arrival of visitors to New York City emerging from the stairways of the subway system. Those without GPS tended to look up, recognize Manhattan landmarks, check on street signs and make their way. Those looking at their GPS followed their on-screen directions, heads down, apparently unaware of their urban setting and its features. As to those GPS systems in cars, they certainly get you from point to point, even if they do not do much for in-car conversation about what's being seen. And if you decide to visit Cape Cod, may I suggest you turn it off in favor of a colorful local map? Whoever inserted the least scenic, most crowded route to my corner of the Cape obviously had no geographic awareness.

Geography's utility certainly made news shortly after the terrible tsunami of December 26, 2004, when the story of a schoolgirl named Tilly Smith made headlines around the world. Tilly was vacationing in Phuket, Thailand with her parents and was on Maikhao Beach when she saw the water suddenly recede into the distance. She remembered what she had just been taught by her geography teacher, Mr. Andrew Kearney at Danes Hill Prep School in Oxshott, south of London: that the deep wave of a tsunami sucks the water off the beach before it returns in a massive wall that inundates the entire shoreline. Tilly alerted her parents and they ran back and forth, warning beachgoers of the danger and urging them to seek shelter on an upper floor of the hotel nearby. About 100 people followed her advice; all survived. Of those who stayed behind, none did. Britain's tabloids declared Tilly to be "The Angel of Phuket," but give some credit to that geography teacher who obviously had the attention of his students.

Okay, you might say. As an everyday tool to make life a bit more predictable and efficient, and as an occasional environmental alert, geography has its uses. But does that make it important in a general sense?

Consider this: a general public not exposed to a good grounding in geography can be easily confused, even misled, as they follow the sometimes contradictory results of ongoing scientific research. Even today, despite the best efforts of the National Geographic Society and its allies, an American student might go from kindergarten through graduate school without ever taking a single course in geography—let alone a fairly complete program. (That's not true in any other developed country, nor in most developing ones. Geography's status is quite different in Britain, Germany, France, and such countries as Brazil, Nigeria, and India.) Some of us recall (and certain newspaper columnists remind us of) scientific studies published in the 1960s that forecast imminent glaciation. Bitter, lengthy winters were driving people who could afford to do so out of the Midwest and other northerly locales towards what came to be known as the "sunbelt." Newspapers carried scientists' dire warnings about ever-shorter summers and even tougher winters for the immediate future. But before the end of the century, a reversal was underway, and the warming phase now in progress entailed forecasts of torrid temperatures, longer summers, rising sea levels and environmental extremes.

There's nothing like early and sustained geographic education to make sense of such apparent contradictions. That cooling phase in the mid-twentieth century had causes that were partly natural—climate change is a permanent feature of our planet—and partly caused by factory emanations whose effect was to reflect the sun's radiation back into space. The warming phase of the present also results from a combination of causes, but now the human factor contributes to heating, not cooling. Not only is the industrial contribution quite different because of changing technologies, but the volume of pollution poured into the atmosphere is far larger: the population explosion and industrial expansion of the past century was just getting underway a half century ago. To get a picture of the reasons behind the apparent contradiction, it helps to understand the workings of nature as well as the growing impact of humanity on our planet—the combination of topics that defines introductory geography and gets students ready for specialization later in their education.

When I talk about this issue on the public-lecture circuit, someone in the audience is likely to challenge my point about the state of geographic knowledge. It may be bad, goes the argument, but don't worry, our leaders know what geography they need to know. They deal with the world at large on a daily basis, and they're sure to be adequately informed and prepared.

Well, maybe, although I wonder about those leaders who come from elite universities that do not offer any geography as part of their undergraduate or graduate curricula. Do you suppose that, if former defense secretary Robert McNamara had been able to take just one course in basic regional or human geography at his alma mater (Harvard), his perspective on Southeast Asia in general and Vietnam in particular might have been different? I would like to think so, but Harvard University has not offered geography to its students for about a half century. The cost to the country may be greater than we can imagine.

WHY GEOGRAPHY MATTERS TO ME . . .

"I [am] convinced that geography is the foundation of all . . . When I begin work on a new area . . . I invariably start with the best geography I can find. This takes precedence over everything else, even history, because I need to ground myself in the fundamentals which have governed and in a sense limited human development . . . If I wanted to make myself indispensable to my society, I would devote eight to ten years to the real mastery of one of the world's major regions. I would learn the languages, the religions, the customs, the value systems, the history, the nationalisms, and above all the geography."

—James Michener in *Social Education,* 1970

"The study of geography has been debated by Americans for many years . . . whether or not it is appropriate for Harvard to teach geography, it is certainly vital knowledge for our citizens and our students. With all [the] deficiencies in our education, it should not be surprising that so many Americans and so many students know so little about geography. Like it or not, the policies, indeed the future, of the U.S. [will] be influenced by many events that happen abroad, and by peoples of other nations, and even by the physical geography of other parts of the world. The world is shrinking, and . . . more and more events impact, or will impact, the United States . . . All of this starts with geography."

—Caspar Weinberger in *Forbes Magazine,* 1989

"During my time as Secretary of State, I witnessed firsthand how important it was that Americans understood geography and the world around them. Since then, as countries have become ever more interconnected, that need has grown."

—James A. Baker III, U.S. Secretary of State (1989–1992), quoted in the AAG *Newsletter,* Vol. 46, No. 10, November, 2011

"Geography played a leading role in nearly every policy decision I was involved in as Secretary of State. Young Americans with an understanding of peoples, places, and cultures have a clear advantage in today's rapidly-changing global economy . . ."

—Madeleine K. Albright, U.S. Secretary of State (1997–2001), quoted in the AAG *Newsletter,* Vol. 46, No. 10, November, 2011

THEY MAJORED IN GEOGRAPHY . . .
Prince William (Duke of Cambridge)
Michael Jordan (NBA star, Chicago Bulls)
Augusto Pinochet (Military Ruler of Chile)

As to our leaders knowing the map they must navigate, consider this little incident in President Nixon's Oval Office, as described by another Harvard figure, former secretary of state Henry Kissinger, in his book *Years of Renewal*:

> As part of some U.N. celebration, the Prime Minister of Mauritius had been invited to Washington. Mauritius is a subtropical island located in the Indian Ocean . . . it enjoys plenty of rainfall and a verdant agriculture. Its relations with the United States were excellent. Somehow my staff confused Mauritius with Mauritania, an arid desert state in West Africa that had broken diplomatic relations with us in 1967 as an act of solidarity with its Muslim brethren in the aftermath of the Middle East War.
>
> This misconception produced an extraordinary dialogue. Coming straight to the point, Nixon suggested that the time had come to restore diplomatic relations between the United States and Mauritius. This, he noted, would permit resumption of American aid, and one of its benefits might be assistance in dry farming, in which, Nixon maintained, the United States had special capabilities. The stunned visitor, who had come on a goodwill mission from a country with, if anything, excessive rainfall, tried to shift to a more promising subject. He enquired whether Nixon was satisfied with the operation of the space tracking station the United States maintained on his island.
>
> Now it was Nixon's turn to be discomfited as he set about frantically writing on his yellow pad. Tearing off a page, he handed me a note that read: "Why the hell do we have a space tracking system in a country with which we do not have diplomatic relations?" (Kissinger, 1999)

So don't be too sure about geographic knowledge in Washington, D.C. It's pretty obvious that we were not well enough acquainted with the physical or cultural geography of Indochina when we blundered (McNamara's word) into the Vietnam War, and I am sure that many of us had doubts about our leaders' knowledge of the regional or human geography of Iraq in the winter of 2003—remember those cheering, grateful crowds that would line the roads? I often cite that old canard about war teaching geography, but in our case we must add a word: belatedly.

Perhaps the most important byproduct of geographic learning, early or belatedly, lies in its role as an antidote to isolationism. Can there be a more crucial objective than this? In our globalizing, ever more inter-connected, still-overpopulated, increasingly competitive, and dangerous world, knowledge is power. The more we know about our planet and its fragile natural environments, about other peoples and cultures, political systems and economies, borders and boundaries, attitudes and aspira-tions, the better prepared we will be for the challenging times ahead.

From this perspective, geography's importance is second to none.

HOW DID IT COME TO THIS?

There's no denying it: for all its putative importance, geography as a school subject and as a university discipline in the United States is, to put it mildly, underrepresented. This wasn't always the case. There was a time when geography was well established as a discipline at Harvard and Yale, when geography was also widely taught in America's schools. During and after the First World War, through the interwar period and again during and after the Second World War, geography was a promi-nent component of American education. In prewar debates, wartime strategy, and postwar reconstruction, geographers played useful, some-times crucial roles. Geographers were the first to bring environmental issues to public attention. They knew about foreign cultures and econo-mies. They had experience with the workings of political boundaries. They produced the maps that helped guide United States policies.

In the 1950s and early 1960s, Americans continued to be well versed in geography. American success during the Second World War had drawn our attention to the outside world as perhaps never before. Maps, at-lases, and globes sold by the millions. The magazine with geography's name on it, *National Geographic*, saw its subscription grow to unprece-dented numbers. University Geography Departments enrolled more stu-dents than they could handle. When President John F. Kennedy launched the Peace Corps, geographers and geography students were quickly ap-pointed as trainers and staffers.

But, as so often happens when social engineers get hold of a system that's working well, the wheels came off. Professional educators thought they had a better idea about how to teach geography: rather than educat-ing students in disciplines such as history, government, and geography, they would teach these subjects in combination. That combination was called social studies. The grand design envisioned a mixture that would give students a well-rounded schooling, a kind of civics for the masses, which implied that school teachers would no longer be educated in the disciplines either. They, too, would study social studies.

Prospective teachers from the School of Education had been among my best and most interested students at the University of Miami during the early 1970s. They registered in large numbers in two courses: World Regional Geography, which was an overview of the geography of the wider world, and Environmental Conservation, a course that was years ahead of its time, and to which even the Department of Biology sent its students. But when the social studies agenda took effect, the student teachers stopped coming. They now had other requirements that precluded their registration in geography.

We geographers knew what this would mean and what it would eventually cost the country. The use and knowledge of maps would dwindle. Environmental awareness would decline. Our international outlook would erode. Our businesspeople, politicians, and others would find themselves at a disadvantage in a rapidly shrinking, ever more interconnected—and competitive—world. Many of us wrote anguished letters to government agencies and elected representatives, to school district leaders and school principals. Fortunately, many private and parochial schools continued to teach geography. But for public education, the die was cast.

REVERSAL OF FORTUNE

This set of educational circumstances in little more than a decade produced exactly what geographers had predicted: an evident and worsening national geographic illiteracy. All of us who were teaching at the time have stories of students' disorientation, some of them amusing, most of them worrisome. By its very name, the catch-all social studies rubric excluded the elementary but crucial physical geography (including basic climatology) topics that had been part of the high-school geography curriculum. This was the one subject in which students got an idea of the importance of understanding human-environment interactions as well as the workings of climate and weather, and it was a huge loss. When these students got to college and enrolled in a first-year geography course, they were at an enormous disadvantage: they simply did not know these basics.

Some university faculties recognized this situation and decided to do something about it. Georgetown University was one of them, and I saw the results firsthand while I was on the faculty of Georgetown's School of Foreign Service from 1990 to 1995. Every incoming student was required to take a course called Map of the Modern World, a one-credit course offered by the noted political geographer Charles Pirtle. In one semester, students were expected to become familiar not only with the layout of the political world, but also with general patterns of geopolitical change, general environmental and climactic conditions, and resource distributions. It was a tall order, but here is what impressed me most: at the end of their four-year degree program, Georgetown students are asked to list the course that pushed their knowledge forward more than any other. Map of the Modern World, a freshman geography course you would think most students had long forgotten, led the rankings year after year. It was a tribute to Charlie Pirtle, to be sure—but it also said something about the relevance of geography in the opinion of these capable students.

Unfortunately the Georgetown remedial model was (and still is) a rarity, not a commonplace. The geographic illiteracy of entering freshmen lowered the level of academic discourse in many an introductory class, and faculty devised various ways of dealing with it. Some professors were, shall we say, more sensitive to students' problems than others, and occasionally stories leaked out about embarrassing moments in the classroom. One of these stories involved a colleague of mine at the University of Miami who liked to start his class by asking students to identify a number of prominent geographic locations on a blank map of the world's countries. The results were always abysmal, and they grew worse as time went on. The good professor would grade the class as a whole and, reportedly with biting sarcasm, would announce the large percentage of participants who could not locate the Pacific Ocean, the Sahara, Mexico, or China.

Early in the fall semester of 1980, the student newspaper, the *Miami Hurricane*, got hold of the test, a summary of test results, and the professor's witty commentary. The paper's front-page story on this tale of "geographic illiteracy" was picked up by the major news media. NBC's *Today* show appeared on campus. ABC's *Good Morning America* invited the principals to New York, but the segment was too brief to throw real light on the problem.

The news, however, had spread throughout the country, and while officials at the University of Miami fretted about what the story might do to the university's reputation, teachers elsewhere tried their own tests on their students. We are all too familiar with the results. At one Midwestern college, only 5 percent of the students could identify Vietnam on a world map. At another college, only 42 percent correctly named Mexico as our southern neighbor. Specialists, including some of the very educators who had helped engineer the demise of school geography, claimed to be "dismayed" at such results. While geographers were not surprised, the question was: how would we reverse this ignominious tide of ignorance?

ENTER THE SOCIETY

Tales of on-campus geographic blindness soon led to newspaper stories of public illiteracy as well. Journalists took to the streets with outline maps of the United States and of the world, asking people at random to identify such features as New York State and the Pacific Ocean and (so it seemed) gleefully reporting the embarrassing tallies. Their stories, however, were usually buried among marginalia.

But then something happened that had the potential to change the picture quite radically. President Reagan, upon arriving in Brasilia, the capital of Brazil, to open an important international conference, pronounced himself pleased to be in . . . Bolivia. This caused quite a stir in Brazil, and his faux pas made the front page of *USA Today*, which busied itself identifying similar gaffes by other politicians. Now geographic illiteracy suddenly was headline news, and the television networks fell over themselves covering it. One of them, ABC-TV, called the University of Miami, which relayed the call to me at a hotel in Baltimore where I was attending a meeting. That call led to my first appearance on *Good Morning America*, and the response to my segment (from the Netherlands) generated a week-long geography series a few months later and my six-year appointment to the *GMA* staff as geography editor subsequently.

But it would take more than the support of *GMA*'s perceptive executive producer, Jack Reilly, to make a real dent in our national geographic illiteracy. As it happened, however, I had a parallel opportunity through my appointment as an editor at the National Geographic Society in 1984.

In 1980 I had had the good fortune of being invited to join the Society's Committee for Research and Exploration, and I began almost immediately to discuss ways of involving the Society in the campaign. The Society's president, Gilbert M. Grosvenor, was sympathetic to the idea. He seemed to be galvanized by a Society-commissioned Gallup Poll that proved without a doubt that American students had fallen far behind their European and other foreign contemporaries in terms of their geographic knowledge. When I joined the NGS staff full time in 1984 for a six-year editorial term, I was able to help mobilize a crucial alliance.

To most observers, it would have seemed natural for an organization known as the National *Geographic* Society to come to the aid of the discipline. But it was not so simple. For many years, the Society and the discipline had not enjoyed good relations. To the Society and its leadership, professional geographers seemed snobbish, insulated, and often unimaginative. To professional geographers, the Society's popularization of its magazine and the rubric of geography was inappropriate and misleading. "There's precious little geography in *National Geographic*," said my professor at Northwestern University when I arrived there as a graduate student in 1956. "If you're going to subscribe, you'd better have the magazine sent to your home. Not a good idea to see it in your department mailbox here."

That amazed me. In fact, when I was living in Africa during the early 1950s, *National Geographic* was my window to the world, its maps a source of inspiration. I had written its president, Gilbert H. Grosvenor, in 1950 to tell him so. He sent a gracious letter in response, urging me to continue my interest in geography and inviting me to visit the Society's headquarters "if [I] were ever to come to the United States." But as a graduate student, I soon realized that the National Geographic Society and its publications were generally not held in high esteem among "professional" geographers.

Grosvenor's grandson, Gilbert M. Grosvenor, however, was not one to let such bygones get in the way. He launched a massive financial and educational campaign in support of geography at the school level, realizing better than most of us that the schools and their teachers were the key to the future of the discipline. High-school students, he knew, were not coming to college in any numbers intending to major in geography, because they never saw geography as an option when they graduated. The social studies debacle had pretty well depleted the ranks of geography teachers, so the first order of business was to prepare large numbers of teachers to teach a geography curriculum. Since the geography curriculum itself had atrophied, Grosvenor appointed a prominent specialist in geographic education, Christopher "Kit" Salter, to resurrect it. Salter, in consultation with the half dozen or so geographers on the Society's staff, developed a spatially and environmentally based framework that would come to be known as the "Five Themes" of geography. In 1986 the Society printed several million copies of an annotated map in full color titled "Maps, The Landscape, and Fundamental Themes in Geography," providing every school in the country with as large a supply as needed.

Meanwhile, Salter under NGS auspices organized a nationwide network of so-called Geographic Alliances representing every State in the Union. These alliances consisted of geography teachers supported by the Society in various ways. Representatives of each alliance were invited to Society headquarters in Washington, D.C. for instruction in geographic education; they would in turn assemble teachers in their home States to convey what they had learned. Thus the number of teachers competent to teach geography increased exponentially, as did grass-roots support for the revival of the subject in schools all over the country.

Grosvenor raised significant funding for the project, testified on Capitol Hill on behalf of geography as an essential component of national education standards, buttonholed politicians, and crisscrossed the country speaking for geography. Not everyone on his staff in Washington was enthused by, or even supportive of, all these efforts, and not all professional geographers ensconced in their academic departments appreciated what he did. But the leadership of the Association of American Geographers had the good sense to extend formal recognition to him for a campaign that closed the book on old, painful disharmony between Society and discipline.

DON'T KNOW MUCH ABOUT GEOGRAPHY

So where are we today? I wish I could report that all the foregoing drastically altered the level of exposure of American elementary and high-school students to geography. By the middle of the first decade of this century, the best assessment was that when the Society's campaign began, about 7 percent of American students were getting some geographic education; after nearly 20 years and an estimated investment of about $100 million, that figure was still below 30 percent. Five years later, the picture did not look much brighter. "Don't Know Much About

Geography" was the headline in the *Wall Street Journal* of July 20, 2011, reporting the results of the National Assessment of Educational Progress, also called the Nation's Report Card, that tested how U.S. Students were doing in geography (Banchero, 2011). Only one-third of American fourth-graders could determine distance by using the scale on a map. Less than half of eighth-graders knew that Islam originated in what is today Saudi Arabia. "Geography Report Card Finds Students Lacking," headlined the *New York Times* on the same day, stating that high-school seniors demonstrated the least ability in geography, with only 20 percent found to be proficient or better, compared to 27 percent of eighth-graders and 21 percent of fourth-graders (Hu, 2011).

Why is the picture so bleak? It is not especially encouraging to report that things looked no better in history or civics exams; some analysts were quoted as saying that the social sciences, especially geography, are being pushed out of school curricula because of the intense focus on mathematics and English as required by the legal stipulations of the No Child Left Behind program. But other observers noted that the amount of classroom time allotted to the social sciences had actually increased on average, although geography seemed comparatively disadvantaged. A board member of the National Assessment Governing Board, Shannon Garrison, was quoted in the *Wall Street Journal* as noting that geography in middle and high schools is often an "unclaimed subject," with the responsibility for teaching it frequently "unclear." My colleague Roger Downs, a professor of geography at Pennsylvania State University who for decades has been in the forefront of campaigns to improve geography's status and prospects, expressed concern that "geography's role in the curriculum is limited and, at best, static."

It is dispiriting to contemplate this picture in the context of then-NGS President Gilbert Grosvenor's optimistic address before the National Press Club on July 27, 1988 in which he described the creation of the Society's National Geography Teachers Alliance program, with the goal of training 15,000 geography teachers through annual, month-long, intensive summer institutes at the Society's Washington headquarters and requiring each teacher to offer at least three in-service workshops in their local school districts: "This summer alone we'll have 700 of them in (their) classrooms" (Grosvenor, 1988). A quarter of a century later we are looking at the same inadequacies that impelled Grosvenor to invest so heavily in the nation's geographic literacy needs. The combination of circumstances that causes this situation is complicated, and geography's plight mirrors a larger crisis in K-12 education that is reflected by American students' deteriorating rankings in international tests of the same kind. But geography's challenge is greater in part because geographic illiteracy infects many educators at all levels (including, in my experience, some college and university deans—this is no K-12 monopoly) through no fault of their own: they came to their jobs without formal education in the subject and have a vague view of its role and importance. In truth, the National Geographic Society's Alliance program was a drop in a very large bucket—a welcome and well-intentioned one, but an effort that did more to prove how hard the task would be than to achieve the goals it set.

This is not to suggest that the NGS campaign has borne no fruit at all. Undoubtedly the situation would be significantly worse than it is had the Society's initiatives and programs not been mobilized; the Alliance teachers brought geography to the attention of students some of whom chose geography as their college or university major, an uptick noted by registrars. The Society's media presence put geography on front pages and in network television programs. Geography's presence in the venerable Magazine is far stronger than it used to be, and its articles and maps form a poster for the discipline in the public eye. To quote professor Downs: "As the economic and cultural forces of globalization and the impacts of global environmental change are felt by everybody everywhere, the case for geography seems both obvious and inescapable." Yes, but in making the case we have a long way to go.

WILL GEOGRAPHY BE HISTORY?

Some of my colleagues take a dim view of the future of geography as a discipline. Yes, the United States Congress endorsed the establishment of National Geography Week every November, and the winner of the annual National Geography Bee, modeled on the famous spelling bee, gets television coverage every spring. National newspapers and network media are paying more attention to geography.

But against these promising developments in the public arena stand some worrisome negatives, two in particular. Ours is a history-obsessed culture. From archeology to geology and paleontology to linguistics, we tend to focus on the temporal. In higher education, spatial science gets short shrift just as geography still does at the school level. To Americans it is inconceivable that a university or college, whether prestigious or

unpretentious, could exist without a history department. No basic curriculum, whether at Harvard or at a Midwest community college, would
exclude a history component. The same cannot be said for geography.

And professional geographers, as we have noted, are divided on the
substance of their discipline. It's probably a healthy debate, it isn't the
first time, and it goes on in other disciplines, too. But it can be confusing
to college and university administrators who read our scholarly journals
and aren't sure just what our consensus is. History, anthropology, and
biology are more clearly defined—they think.

I take a fairly Neanderthal view of this issue. Our basic, common
ground, I feel, lies in regional geography, human-cultural geography,
and physical (environmental) geography, along with the analytical tools
students will need as they begin to specialize even at the undergraduate
level, ranging from statistical analysis to Geographic Information Systems. Beyond this, the tie that binds us—but need not constrain those
who go off in other directions—is the spatial perspective and spatial
analysis. To those who doubt geography's disciplinary future I say that
our great opportunity lies at the interface of environment and humanity. We have been at this for the better part of a century and we were
ahead of our contemporaries for much of that time. We should reclaim
our position.

As to geography becoming history, I must tell you that I admire and
envy the way historians have made their case to the general public as
well as academically. Every time I turn on my television I seem to find
some "presidential historian" commenting on good deeds and misdeeds
of former presidents. And I agree: it is true that we should be reminded
now and then of what President Nixon knew about Watergate and when
he knew it. *When*, after all, is history's key question. But more recently
we had a president who evoked the question: what did the president do
and *where* did he do it? That's geography! We need a presidential geographer! My proposals to this effect have, for some reason, been ignored
by the networks.

Seriously, we professional geographers have not done an adequate
job of informing the general public of what it is we do—and why geography matters. We may not be alone in that respect; scientists in other
disciplines also contribute to the public perception that scientific research
tends to be conducted behind the walls of academia's "ivory tower."
Without having reliable evidence at hand, I surmise that only a very small
percentage of scientists feel comfortable in the public arena, confident

enough to explain to ordinary people why what is being studied is important and relevant to those paying the bills as taxpayers or donors. But among those very few are scholars whose impact on the public—and on their disciplines—has been exemplary. The astronomer Carl Sagan, whose research focused on the physics and chemistry of planetary surfaces and atmospheres, did much to draw public attention to cosmology at a time when space probes were opening new scientific horizons. His interest in the origins of life on Earth and in the search for extraterrestrial intelligence fired the public imagination, and through a series of books and a highly successful television program, Sagan popularized not only cosmology but also simultaneous advances being made in evolutionary theory and neurophysiology. How many young students he attracted to these and related specializations will never be known, but many of them now work in America's public and private space programs.

Some readers may judge that it is rather easier to get the public excited over cosmology and space exploration than over geography. Indeed, the talented *New York Times* science writer John N. Wilford, invited to address a plenary session at the 2001 meeting of the Association of American Geographers, opined that geographers have "done a poor job of speaking the popular language, of conveying in simple and direct terms what is important about their work" (Wilford, 2001). But when geographers have the opportunity, they tend to find a very interested and receptive public audience, because it is not difficult to relate geography to immediate and daily concerns that affect us all, from climate change to the rise of China and from globalization to terrorism. In the process, it is always gratifying to hear from a listener or viewer who says that the geographic perspective on old and seemingly intractable problems is new and exciting. It's worth the effort.

GEOGRAPHIC LITERACY AND NATIONAL SECURITY

Geographic knowledge is a crucial ingredient of our national security. We have crossed the threshold to a century that will witness massive environmental change, major population shifts, recurrent civilizational conflicts, China's emergence as a geopolitical as well as an economic superpower, unifying Europe's transformation into a major player on the international stage—among other developments yet unforeseen. Among my colleagues are geographers who conduct research on the likelihood of coming energy crises and how to forestall them, on the risks of WMD (Weapons of Mass Destruction) dissemination and how to mitigate them, on the impact of global climate change in especially vulnerable areas and how to confront it. These are serious issues indeed, and while geographic knowledge by itself cannot solve them, they will not be effectively approached without it. WMD diffusion, for example, is driven by technology as well as ideology. The technology is the stuff of other disciplines, but ideology has significant geographic ramifications. Extremism of the kind that propelled the Taliban movement to power in Afghanistan from its bases in mountainous and remote western Pakistan tends to fester in isolated locales, and there is nothing uniquely Islamic about this. States that fail, at dreadful cost to their inhabitants, tend to lie segregated from the mainstreams of global interaction and exchange. From Somalia to Afghanistan, from Cambodia to Liberia, from Myanmar to North Korea, their peoples pay a terrible price.

Geography is a superb antidote to isolationism and provincialism. Some specialists in geographic education argue that our persistent national geographic illiteracy results from our own "splendid" isolation between two oceans and two nations, but we are learning that this spatial solitude means little in a fast-globalizing world. During the Vietnam War, there were politicians who advocated "bombing the North back to the Stone Age," and the United States had the power to do so. What the United States was unable to do was to persuade tens of millions of Vietnamese to change their ideology. More recently in Iraq, military intervention proceeded quickly and efficiently, leading to premature assertions that the war was won. But the real war, for Iraqis' hearts and minds, still lay ahead and entailed a costly insurgency that devastated the country's heartland, was countered by an uneasy alliance between invaders and former tribal enemies, and sowed the seeds for post-occupation violence. The United States and its allies had equipment and ordinance, but could not forestall the sectarian strife that accompanied and followed the withdrawal of American forces from Iraq in December 2011. Too few Americans knew the region, spoke the languages, understood the customs and rhythms of life, comprehended the depth of feelings.

And too few Americans understood the geographic implications of the Iraq intervention. Commenting on "Iraq's Tenuous Post-American Future," L. Paul Bremer, former U.S. Presidential Envoy to Iraq, postulates that "Geography is forever and Iraq lives in a rough neighborhood . . ." (Bremer, 2011). How's that again? Geography is *forever*? Tell that to the

GEOGRAPHY AND FOREIGN POLICY

As a professional geographer living in Washington in the 1990s, I dreaded the intermittent appearance of media reports on international surveys that ranked American high-school students near the bottom of the geographic-literacy league. Dinner-party conversation would be spiked with sarcastic commentary ("they couldn't name the Pacific?") and enlivened by amusing stories of adults—some of them politicians and diplomats—embarrassing themselves and their nation in international settings. A repeat favorite concerned President Reagan, who had opened a conference in Brasilia by pronouncing himself pleased to be in Bolivia.

Worse, those reports and anecdotes tended to confirm the public's image of geographic knowledge as equivalent to skill in naming places. It is a useful skill, to be sure, but it has about as much relevance to geographic knowledge as a vocabulary table has to literature. No, geographers were troubled by the decline in geographic literacy in America because we knew it would have foreign policy implications.

WHAT IS LOST WHEN GEOGRAPHIC EDUCATION— AT ALL LEVELS—WITHERS?

What is lost when geographic education—at all levels—withers? Take a comprehensive undergraduate curriculum in the "social" sciences and you will see three recurring perspectives: the temporal (historical), spatial (geographic) and structural (political, economic). Each informs the others, but the spatial perspective is indispensable because it alerts us to the significance of place and location in any analysis of issues ranging from the environmental to the political. That's why geographers tend to reach for their map when they first hear of a major development—such as the intervention in Iraq—and put their Geographic Information Systems to work. But as those dreaded surveys show, even well-educated Americans, on average, are not able to use maps to maximum effect.

Chinese who have transformed their Pacific Rim provinces from backwater to global juggernaut in little more than one generation. Or the occupants of luxury high-rise apartments and villas in Dubai, where modernization meets Islam. Or the residents of Singapore who can remember their city-state's desultory days of stagnation. Or the citizens of the former Soviet Union, who witnessed the disintegration of their political-geographic edifice. No, geography is anything but forever. Start

A second, and crucial, loss involves environmental awareness and responsibility. Geography, alone among the "social" sciences, has a strong physical—that is, natural—dimension. Before geography's decline in American high schools, young students first heard of weather systems and climate change in their geography classes and learned how resource distribution relates to conservation and responsible use.

My geography teacher, Eric de Wilde, raised a question in class in 1948 that kept me thinking forever after: Given the seesaw of ice-age temperature changes, how has history been influenced by climate? From him I learned that we live in an ice age and that we are lucky to experience a brief warm spell between glaciations. Ask the average citizen today what the difference is between an ice age and a glaciation and you are not likely to get a satisfactory answer. Small wonder that politicians can capitalize on public confusion.

So long as we have national leaders who do not adequately know the environmental and cultural geographies of the places they seek to change through American intervention and whose decisions in environmental arenas are insufficiently informed by geographic perspectives, we need to enhance public education in geography. Whether the world likes it or not, the United States has emerged from the 20th century as the world's most powerful state, capable of influencing nations and peoples, lives and livelihoods from pole to pole. That power confers on Americans a responsibility to learn as much as they can about those nations and livelihoods, and for this there is no better vehicle than geography.

The United States and the world will face numerous challenges in the years ahead, among which three will stand out: rapid environmental change, a rising tide of terrorism empowered by weapons of mass destruction and the emergence of China as a superpower on the global stage. To confront these challenges, the American public needs to be the world's best-informed about the factors and forces underlying them and the linkages among them. Geography is the key to understanding these interconnections.

—Washington Examiner, July 22, 2005

with that premise, and you'll get things wrong. Mr. Bremer ends his commentary by arguing that "President Obama made a serious mistake in withdrawing all American forces" from Iraq. Apparently things went so well, you see, while they were there.

As I will suggest in the chapters that follow, challenges loom from a number of directions—the rise of China and its growing power, notions in Moscow of a Greater Russia to encompass parts of the former Soviet empire, the destabilizing weaknesses of Europe, the ascent of India on the regional and global stage, the economic role of a burgeoning Brazil in a competitive world. But how much more does the general public in America know about China today (or India or Brazil) than it (or its leaders) knew about Southeast Asia four decades ago or the Middle East after 9/11?

If there was a way to mobilize it, I would not only reinstate departments of geography in our "elite" universities but also resurrect regional studies in all such departments, old and new, to ensure that, once again, a growing cadre of field-experienced, language-capable, locally connected scholars would populate government, intelligence, and other national agencies whose efforts will be at least as important as high-altitude weapons delivery, satellite imagery, and GIS scrutiny. Geography, unlike its public image, is an entertaining as well as enlightening field, but what follows is also serious—dead serious.

2

READING MAPS AND FACING THREATS

IT IS OFTEN SAID THAT A PICTURE IS WORTH A THOUSAND WORDS. IF that is true, then a map is worth a million, and maybe more. Even at just a glance, a map can reveal what no amount of description can. Maps are the language of geography, often the most direct and effective way to convey grand ideas or complex theories. The mother of all maps is the globe, and no household, especially one with school-age children, should be without one. A globe reminds us of the limits of our terrestrial living space when about 70 percent of its surface is water or ice, and much of the land is mapped as mountains or desert. A globe shows us that the shortest distance between the coterminous United States and China is not across the Pacific Ocean but over Alaska and the Bering Sea. A globe tells us why Northern Hemisphere countries dominate the affairs of the world: most habitable territory lies north of the equator.

This chapter has a dual purpose. Surveys and polls indicate that a significant majority of Americans do not consult maps routinely, but (I can confirm this from personal experience) tend to be fascinated when provided some guidance as to what a map can reveal. In the most general sense, we can divide maps into two categories: representative and thematic. A representative map depicts a particular feature or combination of features on the surface of the Earth as close to reality as the scale will allow. For example, a map of the U.S. Interstate Highway System represents every existing four-lane highway in the country; a map of European cities with 500,000 inhabitants or more represents the available data on urban population in that geographic realm. Every world atlas carries a map of the countries (states) of the world, based on the internationally agreed position of boundaries.

A thematic map, on the other hand, focuses on some physical, cultural or other criterion that may not show up on satellite images (as those highways do). A map of the world that divides global climate into several categories ranging from tropical rainforest to Arctic tundra has an environmental theme. A map of Iraq that shows where Shi'ites, Sunnis, and Kurds have their respective historic domains has a cultural theme. To take this last example, there is no way to draw that map without generalizing: the borders between ethnic-cultural sectors of Iraq's (or any country's) population are almost never sharply defined. That's why thematic maps often lead to debates, and why we should always examine the map's legend and its symbols to find out what the map-maker had in mind. As we will see, cartographers' motives are not invariably noble.

So our first objective is to reflect on map-making and map-reading and examine the ways a map's secrets can be unlocked. As I suggest to my students, please scrutinize a full-page map for at least as long as it would take you to read a page of the book of which it is a part. Try to summarize what it tells you, information that you have not learned from text. I must also report that I do not find anything demeaning or deprecating in asking students to become as familiar as they can with the layout of our world—its geographic realms, regions, countries, even major States and provinces. American and Allied soldiers fought and died in Iraq for nearly a decade, and still only one in seven Americans can identify Iraq on an outline political-geographic map of the world? Three of four cannot locate the Persian Gulf? Such data suggest a widening gap in geographic knowledge at a time when our world is shrinking upon us.

The second purpose of this chapter has to do with the use and misuse of maps. There's nothing like a map to convey a risk, hazard, or threat—yet maps can also be used to deceive and confuse. This takes us into the arena of thematic maps, which can be enormously helpful in clarifying complex issues but must always be carefully scrutinized. But first, a look backward.

FROM CLAY TO COMPUTERS

Cartography, the drawing of maps, has come a long way since ancient Mesopotamians 5,000 years ago scratched grooves in moist clay to represent rivers and fields and let the sun bake it into clay tablets. The evolution of cartography is a stirring story well told by John N. Wilford in *The Mapmakers* (Wilford, 1981), and the saga continues. During the first period of the still-continuing Age of Discovery, explorers, mercenaries, speculators, and adventurers sailed from Europe into the unknown, and those who survived brought back pieces of the great global puzzle for cartographers to fit into their maps.

Magellan and his crew were the first to circumnavigate the world (1519–1522), building on Cabral's impressions of the coast of Brazil and proving the vastness of the Pacific; the Italian Battista Agnese's 1544 map of the world was soon renowned for its beauty as well as its novelty; and the Flemish mathematician and cartographer Gerardus Mercator formulated a grid for the evolving map of the world that allowed navigators to plot a straight-line compass bearing, the Mercator Projection (1569). This was a momentous innovation, and the name Mercator remains famous to this day—as well as another of his inventions, the concept of the atlas (which he named after a Greek titan) as a systematic collection of maps.

It is well to remember, however, that Europeans were not the only map-makers of the time. The Chinese were making maps perhaps 3,000 years ago, and their fleets, larger than anything Europe had floated, were plying Asian and African coasts before Magellan made his epic journey. When Captain Cook traversed the Pacific in the eighteenth century, the local islanders showed him the way through maps made of sticks, fibers, and shells. The ancient Maya and the Inca also made maps. Mapmaking was not a European monopoly.

But the Europeans did collect and assemble the information necessary to create the first representative maps of the entire world rather than just their own realm, and they got better at it as time went on. From Vasco da Gama at sea to Burton and Speke on land, "discoveries" (a word to which, these days, the descendants of the discoverees tend to object) were made at ground or from water level. The explorers and their improving equipment, ranging from compass to sextant and from astrolabe to chronometer, eventually achieved remarkable accuracy and amazing interpretive detail. Nineteenth-century maps tend to represent science, not art, although many were still hand colored for clarity. But the twentieth century witnessed the revolution that would transform cartography and is still under way: the introduction of photography from airplanes, the launching of image-transmitting orbital satellites, and the coming of the computer age.

In the process, the very definition of the term "map" has changed. In traditional works on cartography such as *The Nature of Maps* by Arthur H. Robinson and Barbara B. Petchenik (1976) or P. C. and J. O. Muehrcke's *Map Use* (1997) the authors define a map as "a graphic representation of the milieu" or "any geographical image of environment." But today we see maps of the brain, of human DNA, of ozone holes in the atmosphere, of Mars, of galaxies. High technology, in the words of Stephen S. Hall, "has completely stolen cartography from the purely terrestrial domain" (Hall, 1993).

Well, not entirely. Scientists may use "maps" of chromosomes, regions of the brain that activate when music is heard, or galactic realms where the cosmic action is, but we still use maps for planning a trip, for getting around, for checking on the weather, for getting a sense of where something important in the country or the world is taking place. Unfortunately, surveys show that many Americans are unable to make full use of such traditional maps, even simple ones in commercial road atlases. They have trouble dealing with the standard properties of ordinary maps, such as scale, orientation (direction), and symbols. They find it difficult to relate the legends of maps to the contents of the maps themselves. It is also easy to be confused by the effects of certain map *projections*, for example the Mercator map, which distorts the size and shape of landmasses and oceans almost comically. Take a look at the world on a Mercator projection, and you see that Greenland (for example) appears larger than South America when, in fact, South America is eight times as large as Greenland (Fig. 2-1).

Although the world of cartography is rapidly changing as a result of technological innovations, it's still useful to be able to interpret what printed maps show. Anyone who sells ground-based real estate is still subjected to the requirement of a "current" survey, a costly process that will bring surveyors to your property who spend a day or more finding markers, sighting corners, measuring distances and, eventually, producing a survey map of it showing dimensions, prominent features and, if any, encroachments. If the property is bought and then sold a few months later, that "old" survey won't suffice and it has to be done all over again. It's worth looking closely at any survey: it is the buyer's responsibility to be aware of what it shows.

Whether a survey or a tourist map, a subway diagram or an atlas map, any cartographic representation of the real world can challenge the user. It always helps to assess the purpose and limitations of a map before

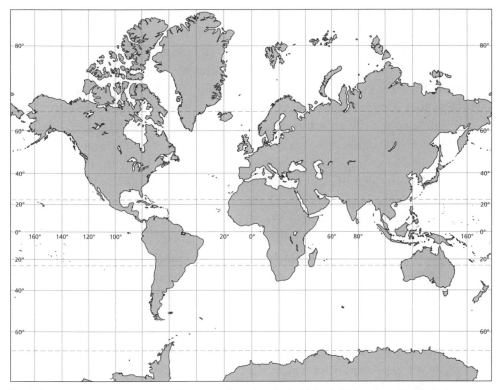

Fig. 2-1. Mercator's projection and its huge size distortions in the higher latitudes. Greenland looks to be larger than South America and on a par with Africa, when in fact South America is eight times larger.

using it. Does that diagram you picked up at the mall entrance have a scale, so you can judge how far you have to walk (and how long it will take) to get to Macy's? Does the city map you are navigating on have a north arrow, so you can gauge direction as you go? And what do those bright green patches on that State map of California mean? Let's have a brief look at the sometimes-hidden keys to a map's utility.

SCALE: HOW FAR IS IT?

There is no escaping it: any map, if it encompasses a section (or all) of the Earth's surface, must represent a rounded surface on a flat piece of paper or a flat screen. The larger the segment of the globe thus represented, the greater the problem. It hardly matters when it comes to a map of a small town, where the curvature of the Earth has little effect. But a map of the entire United States needs substantial "flattening," and a map of the entire world requires some complicated manipulation to avoid crippling distortion.

The larger the area represented on our flat piece of paper, the smaller the scale of the map and the less the detail it can display. This is one of those apparent contradictions in the language of geography: you would think that a map of a whole continent is a large-scale map, because it covers such a large area, but in fact it is a small-scale map. A page-size map of a city block or suburban street where you live covers a small area, yet it is a large-scale map. At this large scale, you can show individual houses, streets, and sidewalks. Most of this detail would be lost, however, if the page were to contain a map of an entire city. Now the scale would be smaller, and only major urban areas and arteries could be shown. Put an entire state on the page, and the city becomes little more than an irregularly shaped patch. In turn, the state becomes just an outline if the page must contain the whole country, with very little detail possible.

Each time, in the example above, we made the map's scale smaller to accommodate an ever larger area on our page. Thus, in addition to the legend and its symbols, we should examine the scale of any map we read. Our expectations of what the map can tell us are based in part on the scale to which it is drawn.

Why is scale smaller when the area represented becomes larger? Because scale refers to a ratio: the ratio of a distance or area on a map to the actual, real-world distance or area it represents. To simplify, let us use the number one for the distance on the map. On our map of a city block or suburban street, 1 inch would represent about 200 feet, or 2,400 inches, so the scale would be 1:2,400. This ratio can also be represented as a fraction: 1/2,400. To get the whole city on the page, 1 inch would have to represent 2 miles, or 126,720 inches. Now the ratio (1:126,720) becomes a much smaller fraction: 1/126,720. As maps go, 1:126,720 still is a pretty large-scale map. To get an entire medium-sized state on our page, our scale would have to drop to about 64 miles to the inch, or 1:4,000,000. And for the whole United States, the scale would be 1:40,000,000.

The scale of map, therefore, tells us much about its intended use. When a developer lays out a new subdivision or a planner considers the placement of a new shopping center, large-scale maps are needed. The useful road maps made available by the American Automobile Association and by state tourist offices are at medium scales. Maps of world distributions (of, say, population growth by country) can be presented at small scales. The map's function is the key to its scale.

A good map is likely to display its scale in one of two (sometimes both) ways: as the fraction just discussed, or as a bar graph, usually in kilometers as well as miles. Using this feature apparently is difficult for many map readers, so road maps also show point-to-point distances along highways, for example between exits on interstates. Again, the larger the scale, the more accurate the distances you derive from the bar graph. On a small-scale map of the world, distortion invalidates it for all but the most general impressions.

There is one map on which you can measure distances with complete confidence: a globe. Take a piece of string or tape, and you can measure the shortest distances between places on the planet using the scale provided. That exercise can be interesting as well as troubling. Interesting, because when you lay that string from, say, New York to Beijing or from Los Angeles to Singapore along what geographers call a "great circle" route, that route may not lie where you expected it to. Troubling, because these shortest-distance routes also reveal how close some of our adversaries are to American cities. When the North Koreans began testing rockets and fears rose that such a rocket might deliver a nuclear weapon, strategists on both sides looked at the globe and calculated that Anchorage, Alaska and Honolulu, Hawaii lay just 5,600 and 7,200 kilometers (3,500 and 4,500 miles) respectively from North Korean territory. The technology of war is shrinking the protective cushion of distance.

DIRECTION—WHICH WAY?

A second element displayed by a map has to do with orientation. Foreign visitors to the United States often comment on Americans' good sense of direction. Europeans, many of whom come from the old, maze-like cities of their realm, are not used to directions that say "go four blocks east on 23rd Street and three blocks north on Fifth Avenue." They may be walking in an American city's square or rectangular downtown street pattern, but their awareness of north, south, east, or west is not part of their customary personal navigation. In most European cities, compass directions simply aren't useful when it comes to finding the way. Americans, on the other hand, are brought up with reference to the compass rose. Ask about someone's home suburb, and the first reaction often includes a directional reference: "Rosewood lies about six miles west of the city."

But this does not mean that directions cannot be confusing, even to American map readers. It is convention that north lies at the top of a map, south at the bottom, east to the right, and west to the left. Yet some maps are not aligned this way, and even experienced map readers can be confused if this is not made clear by a prominent "north arrow" pointing in some direction other than upward. So in addition to checking the legend's symbols, a map reader needs to check the map's general orientation.

The compass rose consists of more than the four main directions. Midway between north and east the direction is northeast, and between northeast and east it is east-northeast. These refinements are of use mainly in navigation. Which leads us to a useful extension of the compass rose to our wristwatch. Imagine that you are standing on the deck of a boat, headed in a certain direction. That direction may be west-northwest or south-southeast. You spot a school of dolphins, just off to the right of your course, ahead of the bow. Rather than calling out the compass-rose direction, a quicker directional reference would be "Dolphins at one o'clock!" Using the clock so that you are always moving toward twelve o'clock is a great way to share quick information on the highway. "Elk in the meadow at three o'clock!" will have all faces pointed in the proper direction; "Elk over on the right!" is far less specific. When I took students to Africa on safari, we always practiced this method— often with excellent results when the moment of observation was brief.

One final point involving direction. Why is it that the world is always represented in such a way that Europe and Asia and North America are at the top, and Australia and Antarctica are at the bottom? Again, this is a matter of convention. While it is logical to use the Earth's two poles as top and bottom of any world map, nothing in nature specifies that the North Pole should be at the top of the map and the South Pole at the bottom. What is now universal practice developed from the work of the earliest cartographers, who lived in the Northern Hemisphere and who started at the top of their page. Most of what was to be discovered turned out to lie to the east, west, and south of their abodes, and so Africa, South America, and Australia came to occupy the bottom half of the evolving world map. It has been that way ever since, except in Australia and New Zealand. There they pointedly draw maps that put "Down Under" on top of the world.

LEGENDS AND SYMBOLS

Getting the most out of reading a map involves interpreting the symbols that make many maps look dauntingly complicated. These symbols range from simple dots or circles to mark the location of towns and cities on small-scale maps to terrain representation by means of contour lines on larger-scale maps. The United States Geological Survey (USGS) for a very long time has published a set of Quadrangle sheets in the 7.5 Minute Series covering all of the United States, and if you have not seen the map that covers your home area, you are in for a pleasant surprise, whether you live in a city or in the countryside, in flat or mountainous areas. These USGS maps are not in fact geologic maps, but surface maps: they show in remarkable detail the slopes and streams, roads and paths, forests and lakes, towns and farms, and virtually all else in the natural and cultural landscape. Some even show individual houses. Like a good book, such a map is hard to put down once you have started "reading" it.

USGS and other relatively large-scale maps do require some study of legend and symbol. They show forested areas, generally built-up areas, the location of major electrical overhead transmission lines as well as oil and gas pipelines, power stations and railroad stations, bridges and beaches. Some symbols are obvious, others take some getting used to. But that is the case with all intricate maps that display lots of information.

A major challenge for cartographers is the depiction of hills and valleys, slopes and flatlands collectively called the *topography*. This can be done in various ways. One is to create an image of sunlight and shadow so that wrinkles of the topography are alternately lit and shaded, creating a visual representation of the terrain. Another, technically more accurate way is to draw contour lines, as is done on the USGS Quadrangle sheets (Fig. 2-2). A contour line connects all points that lie at the same elevation. A round hill rising above a plain, therefore, would appear on the map as a set of concentric circles, the largest at the base and the smallest near the crest. When the contour lines are bunched closely together, the hill's slope is steep; if they lie farther apart, the slope is gentler. Contour lines can represent scarps, hollows, valleys, and ridges of the local topography. At a glance, they reveal whether the relief in the mapped area (the vertical distance between high and low points) is great or small: a "busy" contour map means lots of high relief.

In the United States, contour lines are still measured in feet; in most of the rest of the world, where metric measures prevail, they are in meters.

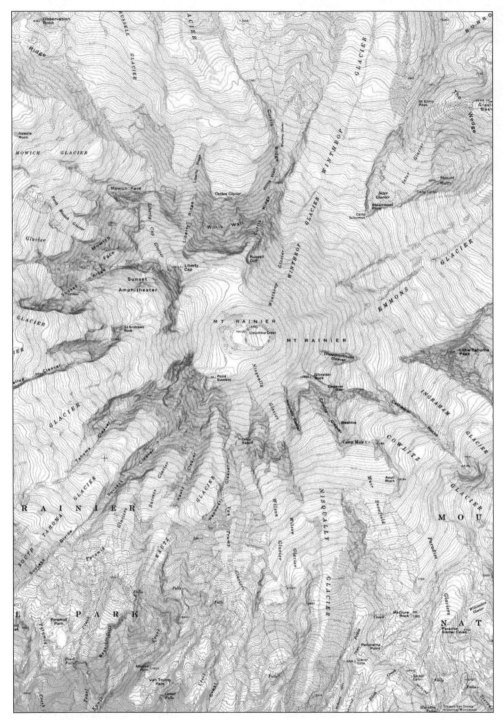

Fig. 2-2. The United States Geological Survey (USGS) map of Mount Rainier, a sample of the superb hand-drawn cartography in this series. It's worth looking at this with a magnifying glass.

One reason the United States had been slow in this global conversion has to do not only with long-term cartographic practice, but with something more far reaching: the Ordinance of 1785. According to this decree, land north of the Ohio River and west of Pennsylvania was laid out in a vast township-and-range system based on six-mile "township" squares along parallel, east-west baselines before it was opened to purchase by farmers. Eventually the system allocated most of the land between the Appalachians and the Rockies, creating the rectangular cultural landscape familiar to anyone who has flown over or driven through Midwestern States. For more than two centuries all the titles and other legal documents pertaining to this vast expanse of the United States have been expressed in English measures; converting it all from miles and acres into kilometers and hectares would be impractical. So nonmetric practice continues here, grooved ineradicably into the soil.

On thematic maps, symbols are the key to interpretation and utility. The maps created by the National Geographic Society, often a combination of representative and thematic, may have as many as a dozen or more symbols in the legend, ranging from the physical (glacier, swamp, dune) to the artificial (canal, airport, railway). For nearly a century, National Geographic Magazine has included such maps, folded to fit, in the journal's issues, a huge bonus for readers and a boon for teachers and travelers. But not all of the maps made by what is now called National Geographic Maps become part of the Magazine. In 2009, for example, at the height of the Allied campaign in Afghanistan, National Geographic Maps published a superb map of that embattled country, not part of any issue of the Magazine but available separately from its headquarters. That map is a truly indispensable guide to the problematic geography not only of Afghanistan but also neighboring Pakistan (and yes, Abbottabad, where Osama Bin Laden was found and killed, is on it). Provinces whose names we would come to know (Kunduz, Kandahar, Helmand) are clearly mapped, and you can find such crucial geographic features as the Khyber Pass, the Vakhan Corridor, the Hindu Kush, Pakistan's part of Kashmir, and the Federally Administered Tribal Areas where Osama was long suspected to be hiding.

This map of Afghanistan displays no information on ethnic or cultural divisions; given the level of other detail, this would have been impractical. But you can tell that the cartographers thought about it. In the center of Afghanistan there is a boldface name that appears to represent neither a province nor any other administrative division, and it is the only designation on the entire map with that level of boldface print. What impelled the mapmakers to make this exception? The answer appears to be: the Hazara are a unique component of Afghanistan's ethnic-cultural mosaic, the only Shi'ite people in a dominantly Sunni country, Dari (Farsi) speakers in a dominantly Pashto-speaking nation. Reason enough, the cartographers must have thought, to put the "domain of the Hazara"—Hazarajat—on a map that does *not* show the majority Pushtuns (Pashtuns), Tajiks, or other ethnic entities in the population.

Even on maps much simpler than those of National Geographic, it is worth scrutinizing the legend. It is possible to miss all kinds of interesting detail without doing so.

MAP PROJECTIONS

This brings us to the interesting topic of map projections. Earlier I mentioned Mercator and his milestone navigation-friendly projection, without saying exactly what a map "projection" is but pointing out that projections inevitably distort the reality on the ground (Greenland's huge size on the Mercator projection is a case in point). The fact is that you can manipulate map projections to exaggerate, diminish, distort, and otherwise modify any representation of any part of the Earth's surface. Countries can be made to look larger, compared to others, than they really are. Places can be made to look closer to you than they really are. Maps can be used for propaganda purposes, or worse. They can be used to stimulate fear, intimidation, aggression, anger, or, at the very least, misjudgments among their readers. So, with any map as with the written word, reader beware!

For centuries mapmakers have grappled with the problem of representing our spherical Earth on a flat surface. To get the job done, they constructed an imaginary grid around the planet, using the poles of rotation and the globe-bisecting equator as their starting points. Since a full circle has 360 degrees, the Earth is divided, pole to pole, by meridians (Fig. 2-3). Of course a starting meridian, or prime meridian, was needed, the zero-degree meridian. This decision was made when the British Empire was at its zenith, and so, not surprisingly, the prime meridian was established as the line of *longitude* running through the Greenwich Observatory near London. That turned out to be a fortunate choice, because the 180-degree line, which would divide the globe into Western

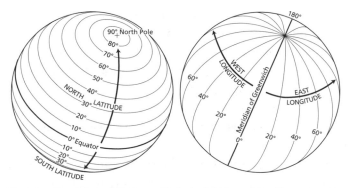

Fig. 2-3.   The global grid in lines of latitude (left) from 0 (the Equator) to 90 (Poles North and South) and longitude (right) from 0 at Greenwich, England to 180 across the Pacific Ocean.

and Eastern Hemispheres, lay on the opposite side of the world from London—right in the middle of the Pacific Ocean.

The meridians are the "vertical" lines of the Earth's grid; they converge on the poles and are farthest apart at the equator. The "horizontal" lines, called parallels, meet nowhere: they form equidistant rings around the globe, starting at the equator. These parallels, too, are numbered by degree: the equator, being neither north nor south, is zero degrees *latitude*. Then the numbers go up. Mexico City lies just below 20 degrees north latitude; Madrid, Spain, just above 40; St. Petersburg, Russia, at 60. Not much action north of this: even the capitals of Iceland and Greenland lie below the Arctic Circle (66°30′ N). The North Pole, of course, lies at 90 degrees, the highest latitudinal point on Earth.

So why is a globe or map grid called a projection? Because that's exactly what it is. You can imagine this by considering an open wire grid with a light bulb at the center and a cylinder of paper wrapped around it. The parallels and meridians would throw shadows onto the paper, creating a kind of projection—not a usable one, but now it becomes a matter of manipulating the light source or the paper. If you put a neon tube inside, pole to pole, you would get something close to the Mercator projection. If you made a conical "hat" of the paper and put it on the Northern Hemisphere, rather than a cylinder all around the globe, you would get less distortion—but only half the world. Indeed, some projections are called "cylindrical" and others "conical." If you want to produce a comparatively low-distortion map of the United States, you would use a conical projection. Projections designed to minimize the distortion

of shape are easier to devise for limited latitudinal areas like the coterminous United States than for the world as a whole, but some remarkable,
inventive projections have been created nevertheless. By "interrupting"
such projections in the oceans, where losing parts of the depicted surface matters less, the shape of the continents is remarkably preserved
(see, for example, Fig. 3-2, pp. 82–83). Projections designed to keep continents and countries as close to their shape and size as possible are called
equal-area projections.

To identify what geographers call the absolute location of a place (as
opposed to relative location, about which more later), we must state the
degrees, minutes, and seconds north or south latitude and east or west
longitude. That seems cumbersome at first, but when you get used to it
you can quickly pinpoint the remotest spot on the globe (atlas indexes
provide these data for hundreds of thousands of locations). Importantly,
no two places have exactly the same location. The recent invention of
Global Positioning System (GPS) technology now makes it possible for
scientists (for example, archeologists) to record the location of a discovery, leave it, and relocate it even if a sandstorm has changed the terrain
unrecognizably. From submerged wrecks at sea to cave entrances on
land, from a single dwelling in an isolated village to a gravesite in an
overgrown valley, GPS equipment records coordinates that are unique
to each. This is one of those "how did we ever do without it" inventions
that have radically changed scientific research over the past few decades.

To return to the global grid of parallels and meridians, Figure 2-3
suggests an important reality: latitude and longitude lines cross at right
angles. Long before the modern grid was laid out, Mercator realized the
implications of this, creating his navigator's chart when much of the
world was still to be "discovered." What Mercator probably did not foresee was that his projection, once all lands had been identified and mapped,
would be used for other purposes as well. The Russians, Europeans, and
Canadians loved it. Leaders and teachers in midlatitude countries liked
a map that boosted the size of their homelands.

For nearly a century, the National Geographic Society often used the
Mercator projection to display world political changes. Then, during the
1980s, the Society's leaders decided to adopt a different projection as
their standard world map, a projection constructed by the American geographer Arthur Robinson (a number of other less-distorting projections
were also in use at the time). Then editor of the Society's scientific journal, called *National Geographic Research*, I was invited to the news confer-

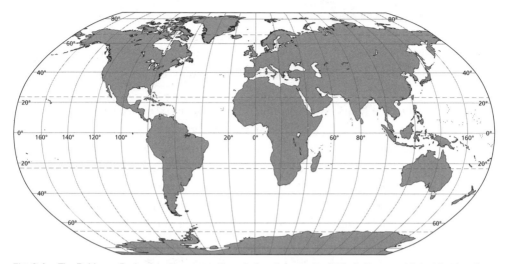

Fig. 2-4.   The Robinson Projection, like many others before it, keeps the latitude lines equidistant but bends the meridians, greatly reducing the size and shape distortion of the Mercator Projection

ence where the change from Mercator to Robinson was announced (Fig. 2-4). When it came time for questions, a reporter from a local news organization rose and asked, "Why has it taken you so long to make this change? It seems to me that the old map reduces the size of Africa and other tropical areas, and it's a kind of cartographic imperialism!" She had a point: the Mercator projection gives not only the United States but also the former colonial countries a huge boost in size. But, as the Society's representative said in response, the biggest "loser" from Mercator to Robinson was the Soviet Union. It "shrank" by as much as 47 percent!

MANIPULATING MAPS

By bending the Earth's grid lines in certain ways, therefore, we can make continents or countries look bigger than they really are, or smaller, or shaped differently. Cartographic deception, intended or accidental, is more common than you might think. The misuse of maps has a long history. The nazis were masters at it. The communists could match them. Advertisers, politicians of many stripes, activists promoting their causes, and others have used misleading maps in pursuit of their objectives.

Sometimes the misuse of maps is unintended, simply a reflection of our general geographic illiteracy. It was painful to receive, a few years ago, a copy of a book titled *Middle East for Dummies* with a map of the region on the cover more than 30 years out of date, still showing the diamond-shaped "Neutral Zone" that once lay between Saudi Arabia and Iraq, just west of Kuwait. Saudi Arabia and Iraq agreed in 1973 to establish a boundary down the middle of this historic remnant, eliminating it from the political landscape. But there it was, on the cover of a 2004 book. Dummies indeed (Davis, 2004).

My colleague Mark Monmonier of Syracuse University's Geography Department has written a series of superb books on many aspects of maps, including deliberate deceptions. His first, *How to Lie with Maps*, points out that while some cartographic distortion is actually intended to deceive or misinform, other maps owe their misleading content to sloppy cartography, poor map design, and inexperience on the part of the mapmakers (Monmonier, 1991). You can confirm this by keeping track of maps in the otherwise punctilious *New York Times*: chances are that a map on Tuesday will be followed by a correction on Wednesday. The problem, as I found out at ABC and NBC, is that maps are often drawn by people whose background is in art, not cartography. Their maps violate all kinds of rules and conventions because they never had any formal training in a geography department's cartography course. Every school of art or design should make a course in introductory cartography mandatory. Every major newspaper should have someone on its staff with some formal knowledge of mapmaking.

There are encouraging indications that this is beginning to happen. The *New York Times* is publishing more sophisticated, thematic maps, based on extensive independent research, to support stories about major international developments. *USA Today* uses color maps to chronicle social change in America. Occasionally newspapers invite geography graduate students with cartographic skills to apply for positions as interns. All this can only enhance their readers' geographic literacy.

As Monmonier points out, there is power in cartography. Imagine that you are opposed to the building of some installation, say a power station, or a waste dump, or a jail, in or near your neighborhood. You plan to distribute a map at the public meeting where this issue is to be discussed. How to enhance the impact of that map? Obviously, the distance from this new facility to the homes most immediately affected by it is a key factor. You can represent this by drawing concentric circles to reflect the anticipated impact—but the map will have much more effect if you make the lines ever thicker as they approach the NIMBY "victims." Now the map conveys the threat far more effectively.

Monmonier's sharp eye spots another, rather less serious, form of deception—a bit of mischief on the part of a cartographer wielding his or her bit of unchecked power. Scanning a 1979 road map of Michigan and northern Ohio, he noted a couple of hamlets with unusual names: Goblu and Beatosu. Take the road past these places, and you will find that they don't exist. The cartographer, obviously a Wolverine fan, must have been carried away by an approaching Michigan-Ohio State football game. Just don't plan to stop for gas at Goblu or Beatosu.

If you have read about hurricanes, tornadoes, earthquakes, volcanic eruptions, floods, or other natural disasters, or perhaps experienced one or more of these terrifying events, you have probably asked yourself where it is safest to live on this continent. In *Cartographies of Danger*, Monmonier warns us that natural disasters aren't all we should be concerned about (Monmonier, 1997). His maps of hazard zones around nuclear power stations, large incinerators, natural-gas pipelines, and other risk-generating facilities in the cultural landscape remind us that we are also a danger to ourselves, especially when we circumvent rules and regulations designed to protect us.

MAPPING AN EVER-CHANGING WORLD

It's just as well that the era of the computer and GIS has made the life of cartographers easier. In decades past I often wondered about the map-making profession and its challenges: boundaries would shift, place names would change, country names would disappear from the map even as new ones emerged. Spend a year working on a print atlas, and before it could be published it would be out of date. I well remember the round-the-clock work that went into the production of the sixth edition of the National Geographic Society's *Atlas of the World* (1992). Someone kept track of the number of changes that went into this new edition and reported that they added up to more than ten thousand—and yet it was out of date when it hit the streets.

But even as we can bring the latest versions of ever-changing maps to our computer screens, maps continue to be printed and undoubtedly will for a long time to come. This is just the latest episode in a sequence of events that began during the age of exploration, when new information about remote locales kept arriving on the drawing tables of European cartographers, keeping them busy revising nonstop. But after the surge of European colonization and before the collapse of the Soviet Union there was a period of relative stability, and map changes were—comparatively—few.

Then, during the period of decolonization, European colonies became independent, new boundaries appeared, new names came and (as in the case of Zaïre) sometimes went. But nothing in recent history matched the decade of the 1990s when, during the aftermath of the breakup of the Soviet Union and the collapse of Yugoslavia, the map of the world changed radically. In the years 1990 to 1999, the United Nations added as many as 32 states to its membership roster. It was a decade of unprecedented global transformation, and it wasn't confined to country names and boundaries. States also made internal changes: South Africa, for example, delimited nine new provinces. India created three new States. China changed from its old spelling to the Pinyin system, so that Peking disappeared from the map and Beijing replaced it; Canton became Guangzhou; Tibet was now Xizang. Hundreds of Chinese names, from provinces to cities to rivers and mountains, had to be changed on every map.

Some of the changes over the past twenty years must have kept cartographers awake at night even with the available GIS technology. Take the case of the Turkish Republic of Northern Cyprus, a reality to be sure—complete with capital city and boundaries—but recognized by only one government (guess which) and no other. How to designate this? Or the so-called Republic of Somaliland, the northernmost arm of this wedge-shaped country where things are going reasonably well and where most of the people want to distance themselves from the chaos in the south?

And while the pace of change seems to have slowed a bit, we're far from stability as yet. During the years of 2000 to 2011, seven more states joined the United Nations, including not only Kiribati and Tuvalu in the Pacific, Timor-Leste (East Timor) in Southeast Asia, Serbia-Montenegro in 2000 (Montenegro seceded from the union and entered the UN in 2006), South Sudan in Africa, and—get this—long-holdout Switzerland in 2002. UN membership is approaching 200 as other hopefuls are waiting in the wings.

One of these hopefuls is Kosovo, another fragment of the Yugoslavian collapse. A Muslim-Albanian former province of Serbia, Kosovo suffered brutal repression during that disintegration, and declared unilateral independence in 2008. The United States supported Kosovo's move

MAPS IN THE MEDIA

Americans may not (yet) be "thinking spatially" to the extent that geographers would like, but there can be no doubt about it: you're seeing more maps in newspapers, magazines, on televisions newscasts, and in the media generally. The *New York Times*, the *Wall Street Journal* and *USA Today* are using maps far more frequently than they used to, and many of these published maps are in color. In recent years the *New York Times* has supported important reports on Iraq and Afghanistan, and on scientific topics in its Tuesday *Science Times* section, with impressive cartography.

There's proof that readers scan these maps pretty carefully. Quite often a *New York Times* map becomes the subject of its famous page-two "Corrections" column after some sharp-eyed reader spots an error. Often the overall map is so effective that one wonders how so obvious an error could have escaped the cartographer's notice. For example, a November 17, 2011 map titled "Battle for the Senate," an analysis of 33 Senate seats up for election in 2012, had a State in the northwest corner of the U.S., bordering Canada, called . . . Oregon!

as did a majority of European and other countries, but Serbia and Russia refused to recognize the new state, and Russia was in a position to block Kosovo's UN membership. So how to map Kosovo?

Thus cartographers not only change or add geographic names, they must also have reason to decide what and what not to put on their maps, and their decisions can produce major controversy. When Yugoslavia collapsed and its "republics" became sovereign states, the Greeks objected vehemently to one of them: Macedonia. The Greeks argued that the name of one of their historic provinces could not simply be appropriated by a neighboring country. The Macedonians said that they had lived in a political entity of that name for generations. The Greeks proposed that Macedonia be called Former Yugoslav Republic Of Macedonia (FYROM). The Macedonians called that ridiculous. So what's a cartographer to do? Whatever goes on the map will stir up anger somewhere. Cartographers cannot please everybody.

My colleague Ki-Suk Lee of Seoul National University reminded me of this when he told me that my use of the name Sea of Japan for the waters between the Korean Peninsula and the islands of Japan was unacceptable to Koreans. For centuries before there was a Japan at all, he wrote me, Koreans had been referring to this body of water as the East Sea, and at a professional meeting soon thereafter he presented an impressive research paper to support his case. At least I was able to compromise in this instance, which is not always possible. My books (including this one) give it both names, East Sea first, Sea of Japan below between brackets. My books, by the way, sell better in South Korea than they do in Japan.

Trying to be even-handed in situations like this can get you into a lot of trouble. Just ask the people at National Geographic Maps, publishers of the eighth edition of the National Geographic *Atlas of the World* (2005). In every previous edition, the water body between Iran and the Arabian Peninsula had been mapped as the Persian Gulf. After the appearance of the seventh edition, however, a professor from a university in the United Arab Emirates wrote to the Society, arguing that this water body is referred to as the Arabian Gulf in his part of the region, and demanded that the next edition of the *Atlas* show it as the Arabian Gulf, not the Persian Gulf. After much deliberation, it was decided to use both names, with Persian Gulf above and Arabian Gulf, in brackets, below. When the *Atlas* was published, the reaction in Iran and among Iranians everywhere was outrage. Tehran prohibited National Geographic researchers or photographers from entering the country and Iranians in the United States launched a wave of criticism—probably the first time in decades that Iranian rulers in Iran and Iranian exiles abroad agreed on anything. And this campaign had serious impact: on the Amazon.com Web site, it lowered the "rating" of the *Atlas* from five points to two; anyone unfamiliar with the background might assume that the *Atlas* was not worth its cost, with serious financial consequences for the publisher of this excellent volume.

In the ninth edition of the *Atlas* (2012) the name Persian Gulf appears in the familiar blue capital letters used to designate water bodies. A small box on the map states that "(it) is refered (*sic*) to by some as the Arabian Gulf." Obviously, the Iranian campaign persuaded the Society to revert to its pre-2005 designation, but another factor came into the picture. The United States Board of Geographic Names (see p. 56), the ultimate arbiter when it comes to official maps published by the U.S. Government, has decided against accommodating "dual" appellations and, after weighing argument and evidence, publishes the approved geographic name. In this case, Persian Gulf is ratified and Arabian Gulf is not.

Obviously, cartographers are not likely to be out of work anytime soon. While mapmaking has entered a new, high-technology age in some countries, maps are still being drawn the old way in much of the world. Traditional and high-tech methods tend to be combined in countries where modernization is going more slowly than it is in the United States, and thematic maps will always require familiarity with design, color gradation, symbolization and other conventions, no matter where they are made. But the entire process is changing, from the sources of data to the role of transmission, and eventually maps will perform functions unimaginable just a few decades ago.

GETTING THE PICTURE

None of this would be happening, at least not yet, were it not for the shock the world got on October 4, 1957, the date that marks the start of the Space Age. On that day, scientists in the service of the Soviet Union launched into Earth's orbit a 184-pound satellite named Sputnik 1, which circled the planet every 96 minutes until it fell back into the atmosphere and burned up in early 1958. In the space race that followed, the Soviet Union and the United States competed for primacy, and in 1959 the Soviets also were the first to hit the Moon with a space probe. But on July 20, 1969 American astronauts were the first humans to walk on the Moon's surface.

In those early days, few foresaw what the space race would yield in the way of what geographers called "remote sensing." Until the 1960s, remote sensing had been confined to aerial imagery, and infrared photography from high-flying aircraft was the ultimate technology. Even when new instruments were invented that could record a much wider range of the electromagnetic spectrum (that is, the electric and magnetic energy emitted at various wavelengths by all objects on the Earth's surface), these instruments could not fulfill their potential.

But then the picture changed. Satellite technology evolved rapidly, and satellites could reach heights aircrafts never could. Soon, artificial satellites carrying ever-more sophisticated remote-sensing instruments orbited the planet, communicating directly with increasingly capable computers to create visual images of the Earth never obtained before.

Among the most productive of these Earth-orbiting satellites was the Geostationary Operational Environmental Satellite (GOES) system belonging to the National Oceanographic and Atmospheric Administration (NOAA). It became possible to place a satellite in a geosynchronous (fixed) orbit, so that it remains stationary above the same point on the Earth's surface. From there, GOES was able to observe the oceans and coasts of the United States, watching for storms and tracking them when they occur. Today, television weather forecasters (some of whom were first trained as geographers) depend heavily on such data. On their televised maps, we can watch weather systems move across the country, a form of animated cartography unheard of just a few decades ago.

Also vital have been four LANDSAT satellites launched between 1972 and 1982 to provide a stream of data about the Earth and its resources. Using a battery of state-of-the-art multispectral scanners and special television cameras, LANDSAT's sensors provided new insights into geologic structures, the expansion of deserts, the shrinking of tropical forests, and even the growth and contraction of algae and other organisms crucial to food chains in the oceans. These satellites enabled the monitoring of world agriculture, forestry, ocean pollution, and a host of other environment-related human activities.

As the capacities of satellite instrumentation improved to the point that it became possible to discern objects as small as cars (and tanks) on the Earth's surface, some satellites were inevitably referred to as "spy" satellites, capable of doing what had required ground reconnaissance before. But countries with a modicum of power and influence were obviously unwilling to have foreign geosynchronous satellites hanging over their territories, so old-fashioned spying continued, as Americans were reminded in 2001 when China forced down a propeller-driven American "surveillance" plane flying just outside its territorial sea, briefly holding the crew and staff hostage on the island of Hainan and causing an international dispute. "Spy, but verify," as Ronald Reagan might have said.

MAPPING SYSTEMICALLY

Coupled with the unprecedented imagery generated by satellite-borne equipment is the equally unparalleled growth of computer versatility, including their graphic performance. Today, the map you see in your favorite magazine may well be drawn by a computer that has been instructed to manipulate information on boundaries, resources, ethnic homelands, or any other spatial feature. That information comes from a

Geographic Information System (GIS). In recent years, geospatial technology has added a crucial dimension to geography.

A GIS system is essentially a collection of computers and programs that combine to collect, record, store, retrieve, analyze, manipulate, and display spatial information on a screen; of course the display can be printed out, so that it can be reproduced in atlases and journals. Because of the huge capacity of today's computers, the amount of information they hold is almost infinite. They allow you to select any area of the world and bring it to the screen at whatever scale you desire. A map of the remaining forest cover in Poland? A map of the oil reserves in Angola? The names and locations of India's new States? It is all there for the asking. This is revolutionary change indeed, but there is more. A GIS allows for a dialogue between the map and the map user; no longer is the map static and unchangeable except through laborious alterations in the subsequent editions of published books and articles. The map user asks for information; the computer guides the user toward answers. This is called interactive mapping, and it is the cornerstone of this latest revolution in cartography. Already, automobiles carry navigation systems that guide the driver toward any address desired, providing voice as well as graphic instructions for the operator to follow. The applications of GIS are limitless.

In the process, as a recent article in the journal *Nature* asserted, "geospatial technologies have changed the face of geography . . . by combining layers of spatially referenced data with remotely sensed aerial or satellite images, high-tech geographers have turned computer mapping into a powerful decision making tool" (Gewin, 2004). The United States government is taking notice: in 2004 the Department of Labor identified geotechnology as one of the three most important emerging and evolving fields, along with nanotechnology and biotechnology. As a result, the job market for geographers is changing as well. "The demand for geospatial skills is growing worldwide, but the job prospects reflect a country's geography, mapping history and even political agenda." In the United States, the focus on homeland security has been one of the many factors driving the job market but, as Gewin points out, there is less demand as yet for remote-sensing expertise in the comparatively small and intensively mapped United Kingdom.

Quoted extensively in this important commentary is the executive director of the Association of American Geographers, Douglas Richardson, who cautions that "although technical skills are important . . . employees need a deep understanding of underlying geographic concepts. It's a mistake to think that these technologies require only technician-oriented functions." This is a crucial observation at a time when the burgeoning of GIS training and development would appear to deplete still more the pool of geography graduates who have research, cultural, and linguistic experience in the field. When concern with United States homeland security focuses on GIS technology to identify, follow, and apprehend terrorist suspects already in the country, we should remember that the roots of Islamic terrorism lie abroad, in the streets, souks, bazaars, and mosques of nations from Morocco to Malaysia where local radicals and American researchers once made mutually instructive contact. The great value of geographers' leadership in GIS should not be undermined by further withdrawal from the real world beyond the computer screen's reach.

NAMING (GEOGRAPHIC) NAMES

The modern era of satellite imagery and GIS technology has brought exquisite detail and unprecedented brilliance to the representation of planetary landscapes on computer screens, atlas pages, and even textbooks. To view the entire Nile Delta from space and to be able to zoom in on local villages and farms, to follow the Amazon River from its tributary sources to the coast, to look at Tokyo as if you're in a spaceship—these are experiences no geography student a half century ago could have. But what is it you're looking at? What is the name of that river flowing from Europe into the Black Sea? How do you spell the name of the capital of Ukraine, seen during a search for nearby Chernobyl? What do the locals in (relatively) newly independent East Timor want to call their country? This, as noted earlier, is no minor matter. Countries like Timor-Leste (that's the answer) or Côte d'Ivoire (please, not Ivory Coast) don't produce the maps and atlases on which their homelands are shown.

In this arena the United States continues to play a crucial role, a kind of global referee that approves geographic names and their spellings. The U.S., through its official publications, validates certain usages and, inevitably, disallows others (that river above has six names along its route but it's called the Danube from source to mouth, which upsets a lot of people along its course). American approval of a particular spelling of the name of a city or mountain range can make locals very unhappy. Even the names of whole countries may cause trouble. Geographic names can be very sensitive issues.

Fortunately, the process of approval is done with the utmost care and consideration by an official committee whose members represent a wide range of expertise and experience. This is the United States Board of Geographic Names, an interagency committee that consists of nine members, each representing a branch of the United States government concerned with or affected by such issues. This nine-member board is divided into two standing committees: the Committee on Foreign Names (four members) and the Committee on Domestic Names (five members).

The Committee on Foreign Names consists of representatives from the Defense Department, the State Department, the Central Intelligence Agency, and the Library of Congress. This group considers primarily changes in, and spelling of, country names, important internal divisions (such as the "republics" inside Russia), and international features—that is, features that extend from one country into another (such as mountain ranges and rivers) and whose names may differ on opposite sides of borders. A small army of staff researchers keeps the committee abreast of changes made within countries, for example, from Leningrad to St. Petersburg, or new spellings, such as Kyyiv instead of Kiev, which are the prerogative of the countries involved.

How does a name change or spelling change become officially approved by the United States? There are various routes. A government may send a formal communication to the United States Secretary of State or to the United States Embassy in the country, requesting affirmation of a change. If consideration of the change falls within the jurisdiction of the Committee on Foreign Names, the file is considered there and either approved or deferred, pending additional information. Eventually, the approved name is codified during a quarterly meeting of the full Board.

Many names, indeed thousands of them, are changed without such requests for approval. Still, the United States needs a consistent form of these for official use. Thus it is the job of the staff researchers to comb government decrees, gazettes, maps, and other sources to secure the necessary information. This information is then collected by the Defense Mapping Agency and submitted to the full Board.

To disseminate the approved names, the Board publishes the *Foreign Names Information Bulletin*, used throughout the government and elsewhere (including atlas makers and map companies) for information on accepted spellings. Only very rarely are the names published in the *Bulletin* not immediately adopted; for cartographers, this is the ultimate source.

MENTAL MAPS

I'm sure you've noticed it: some people seem to have an innate sense of direction. They can find a street or a store with the greatest of ease. They don't miss highway exits and always know where the one-way streets are.

Others are not so lucky. They get in the wrong lane, can't remember on which side of the stadium they parked their cars, lose their way trying to find the home of a dinner host.

Geographers' research has proved that when it comes to the maps in our minds, our *mental maps*, we are not born equal. Just as some people are color blind and others have perfect pitch, the brain's capacity to imagine our activity spatially varies from person to person.

I have some evidence of this in my files. Throughout my nearly 50 years of teaching, I have asked students—not only from America, but from all over the world—to draw maps of their home city, their state, or their country (and sometimes the world) on a blank piece of paper. You would be amazed at the results. Some students can draw, from memory, a remarkably accurate map not only of their city or state, but of any part of the world. Others cannot even draw the barest outlines. Not all of this is a matter of education or exposure to geography. Indications are that our capacity for what the technical people call "spatial cognition" varies quite widely.

Perhaps you have seen those funny postcards they sell in Texas, showing a map of the United States almost completely occupied by the Lone Star State, with all the others squeezed in a narrow band against the coasts and the Canadian border. Well, I have seen this sort of thing in real life—and not as a joke. During the 1960s, when I taught at one of my favorite institutions, Michigan State University, hundreds of students came from Africa to study there. Quite a few of them took my introductory geography class. I always asked my students to draw their mental map of a continent.

Almost always, the American students drew North America, and the African students drew Africa. And almost always I was impressed by the detail African students put on their maps of Africa. But class after class, year after year, I noted something interesting: many Nigerian students drew Africa the way those Texas postcards show the United States. Nigeria would occupy almost all of West Africa and much of Central Africa, too, and the other African counties would lie squashed around Nigeria's perimeter. Nigeria, to be sure, is a large country, and in fact it is the most populous country in all of Africa. But in the minds of quite a few Nigerian students, it was also the Texas of Africa. "Well, Nigerians think big," said a student from Ibadan when I asked him about this. "Let me try drawing the map again." He did, now mindful of his earlier exaggeration. But when he was finished, Nigeria still was about twice the size it should be.

Mental maps can be improved, of course, through the study of geography. My late colleague Thomas Saarinen of the University of Arizona tested students' mental maps throughout the world, with fascinating results. When asked to draw a world map, for example, many students put Europe in the center of it, even when they don't live in Europe and aren't Europeans themselves. That is just one leftover of the educational systems spread worldwide in colonial times.

In general (this should not surprise us), we Americans are rather fuzzier, mental-map-wise, than many of our contemporaries. Does it matter? Geographers think so. As for me, I still remember that day in 1962, when, as a young assistant professor at MSU, I had been invited to join a group of colleagues in the State Department to discuss, with an assistant to Secretary of State for African Affairs G. Mennen Williams, a set of urgent African concerns. But the night before, President Kennedy had appeared on television with a map of Indochina, and in our Washington hotel rooms we had watched his "chalk talk" that revealed the discovery of the Ho Chi Minh Trail. The next morning at State, nobody wanted to talk about Africa. The hot issue was Indochina, and the Trail—and Laos, through which part of it lay. There was much arm waving, numerous proposals and suggestions, but the maps on the wall were of Africa, not Southeast Asia.

"May I just ask," I said as the debate swirled, "can anyone here name the six countries that border Laos? It seems to me that the layout of the region is rather important, given all these ideas."

No one could do it, and worse, no one seemed to care. "That's a waste of time," said one of my colleagues. "If we need to know that, we'll get a map and look at it."

I suggested that if you have a mistaken mental map of a place, you won't know where you're going and that if a whole cadre of decision makers had a vague mental map, we'd be in for bigger trouble than the president had intimated.

In the years since that little incident I've often thought of it and of how little we Americans, on average, really knew about the Indochina where we would wage so costly and bloody a war. Because a mental map is more than a skeleton outline. At its best, it is a store of information, not only about the layout of a place, but also about its components, its dwellings and schools, streets and paths, shrines and markets. It is a map that accrues over a lifetime, the spatial equivalent of our temporal, chronological knowledge and our simultaneous ability to place major events in historical perspective. But our mental maps are not historic, they are current, and they guide and inform our actions and decisions whether a simple excursion for pleasure or a military campaign in a distant land. How clear were they when the Iraq invasion was planned? How clear are they in this century of environmental change, China's rise, and globalization's impact?

PUTTING MAPS TO USE

When geographers are asked to provide an early example of the practical utility of maps in solving real-world problems, we like to go back to the story of Dr. John Snow, a London physician-geographer who lived through several of the dreadful cholera pandemics that ravaged much of the world during the nineteenth century. No one knew for sure how cholera spread, making the disease especially frightening, and many victims died within a week of infection. Dr. Snow had come to believe that contaminated water was to blame, but he had no proof of it. When the pandemic that had begun elsewhere in 1842 reached England, London's densely populated Soho District, near Picadilly Circus, was hard hit. Dr. Snow and his students made a large-scale street map of the area, and marked with a dot the place where each death occurred and where each new case was reported. By the time more than 500 people had died in Soho alone, the map, dated 1854, showed a concentration of victims around an intersection on Broad Street (Fig. 2-5). Now he invoked some geographic reasoning. There were several stores at that intersection, but customers might walk elsewhere if prices a few hundred yards away were lower. But when something is free, they will make a beeline for it, and what was free at this intersection was the water provided by the Broad Street pump. This accounted for the clustering of the dots around the pump, and the link between contaminated water and cholera was confirmed by the map.

That was not the end of the story. Dr. Snow asked city officials to remove the handle from the pump, but they first demurred, saying that

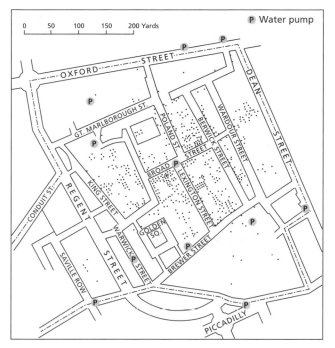

Fig. 2-5. Dr. Snow's now-famous map of a part of London's Soho District, recording the location of new cholera cases as well as deaths resulting from earlier infections. Note the concentration of dots around the water pump at Broad and Lexington.

this would risk a riot in Soho where people were already angry about the casualty toll from cholera. So Dr. Snow and his students did it themselves, pouring lye down the hole for good measure. Soon he had his proof: the number of deaths around the intersection plummeted, new cases dropped even more sharply, and what the map had confirmed was proven beyond a doubt. Now the authorities could advise people to boil their water and to stop worrying about touching each other or inhaling "bad air," two of the suspected causes.

Today, things are not that simple, but maps still help modern epidemiologists trace pandemics, predict future routes of diffusion, and mobilize inoculation campaigns. In this context, GIS technology has transformed the utility of the "medical map": when current information is crucial, it can be in hand in minutes and decisions can be made with far greater confidence. Nevertheless, other technologies create novel challenges. Fast, worldwide travel by jet aircraft can disseminate a group of individuals infected by some dangerous virus in a matter of hours, and once they have dispersed from the airport at their destination, locating them and alerting locals at risk can defeat even the most sophisticated monitoring system.

Dr. Snow's map was a forerunner of countless others in the field of medical geography, but medical geography is only one field under the broad umbrella of human geography. That map pointed to the solution of an existing quandary, but today maps are also used to predict future problems. One especially interesting aspect of this has to do with crime and security. Criminal behavior has a spatial dimension: of course there are random crimes, but many others are planned and executed. Break-ins and thefts ranging from the amateurish to the professional have this in common: the offender starts his or her journey at some point and commits the crime elsewhere. When criminals are apprehended, their records are likely to reveal where they started their "journey to crime" and where they committed it. Make a series of maps of these journeys, and the relevant GIS methodology can not only reveal the way the pattern arose and changed, but also where it is likely to lead, given, say, the expansion of targets in a certain densely populated urban area. This will allow law-enforcement agencies to forecast where security needs to be strengthened—not *after* a crime wave has erupted, but before it happens. This can save money by making police assignments more efficient even as the criminals' risk of arrest rises and security in the anticipated path of the crime wave improves.

As you can imagine, human geography—from medical to criminal and from cultural to economic—has been transformed by such technologies. But history-making maps are not confined to human geography. In some ways the most consequential map ever drawn was the work of a physical geographer, a professional climatologist whose talents reaches far beyond that field. That map, still seen today in physical geography books, geology texts, and even some economic analyses, was first published in a German treatise nearly a century ago (Fig. 2-6). We will take a closer look at it in Chapter 5, and it is likely that every reader of this book became familiar with it in some high school or college course. But when it first appeared it was ridiculed by scientists and the general public alike. Only after Wegener's death did it begin to lead to the discoveries that his hypothetical model proposed.

Wegener's map suggested that, more than a hundred million years ago, the planet's landmasses—the continents—lay united in a giant su-

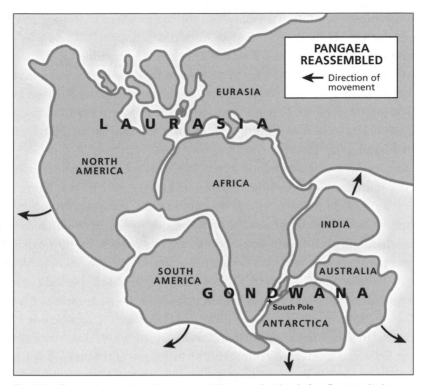

Fig. 2-6. The continents where they were, relative to each other, before Pangaea broke apart and their "drift" began. This map represents the Land Hemisphere about 100 million years ago.

percontinent of which Africa was the core. In a process Wegener called *Continental Drift*, this supercontinent broke apart, with South and North America "drifting" westward and opening up the Atlantic Ocean, Australia moving eastward across what is today the Indian Ocean, and Antarctica southward to its polar position. India, which had adjoined eastern Africa, drifted northeastward and, in a gigantic collision still continuing, pushed up the Himalaya Mountains.

Wegener based his map in part on the apparent "fit" of the landmasses, especially those across the Atlantic Ocean, when they are cut out and placed together. To be sure, he amassed a wealth of other evidence, but he argued that the jigsaw-like match between South America and Africa could not possibly be a matter of chance. After his death, research by geologists yielded a growing inventory of evidence—similar rocks on opposite sides of the Atlantic, geologic structures than began in Africa and continued in South America—but the big problem that dogged Wegener during his lifetime persisted: how did it happen? What was the mechanism that could move whole continents? Decades later, the research Wegener's book had set in motion came up with the answer: the theory of plate tectonics. But by then, the economic as well as the geological implications of Wegener's model made this far more than a scientific issue. His map was a guide to economic geography as well as subsurface geology.

To confirm this, have a look at the Wegener model in the context of what you already know about mineral and energy resources. Note, for example, that South Africa, among the world's leading producers of industrial and gem diamonds, lies adjacent to Argentina—where these same diamonds are found, washed by streams across the non-existent divide before South America separated from Africa. And while oil reserves originated at different times over long geologic periods, some were similarly divided by the supercontinent's fragmentation. Take a look at the known oil reserves of West and Equatorial Africa on one side of the Pacific, and those of northern and eastern South America on the other. Do you see anything remarkable? Where would you look for further discoveries? Is it surprising that Brazil's recent offshore finds happen to lie right across from those of Angola? The map explains why oil was found recently off the coast of Ghana. Reassemble the two continents, and it looks as though additional oil reserves will be found westward along the African coast and eastward along the South American coast. In short: the Wegener map ought to be on the office wall of every commodities broker. It's a guide to what was—and is left to be— discovered.

There was a time when maps were the inventories of discovery, the repositories of the latest information arriving in Europe and later in America from the far reaches of the yet-unknown world. Today maps have totally different functions that range from persuasion to prediction— persuasion, for example, on the back label of a wine bottle where a local vineyard in the US Northwest is placed in the same latitude as Bordeaux of Burgundy and enticing the buyer to try it; prediction, in a time of uncertainty, of trends and developments in places far away that now have huge influence over our daily lives. But, as we will see, maps can also be used for far less constructive purposes. Behind many a political map lies wicked intent.

MAPS OF BAD INTENTIONS

One of my earliest recollections of a map that made me uneasy was during the Second World War, when in my parents' library I ran across a large, glossy magazine in which, right in the middle where the staples were, was a map of Europe with a huge swastika in the center and thick black arrows extending into England, France, Scandinavia, Italy, and Eastern Europe. *Das Neue Deutsche Reich* was the headline above it, and a picture of a scowling Adolf Hitler, arm extended in the nazi salute, was in the top right-hand corner. You didn't have to know any German to know what that headline said. Later during the war, posters began appearing on public buildings and elsewhere with similar maps projecting nazi Germany's imperial designs.

I never forgot that map, even though, when I first saw it, I really did not know enough to understand its implications. If its intent was to frighten, it certainly did so. I'm sure it scared my family even more. After the war, in my geography classes at school, my teacher brought us a map published in Japan, also during the war, that showed much of Pacific Asia as part of a Japanese empire. He explained that governments and regimes sometimes publish maps to proclaim their intentions and that, during the interwar period, some governments even printed stamps that falsified boundaries with neighbors—subtle hints of disputes to come.

Many years later I learned a practical lesson to this effect. A colleague at the University of Baghdad had been sending me, via a brother living in Denmark, maps being published by the regime ruling Iraq. In mid-1990 one of these drew my rapt attention because an apparently official map of Iraq showed Kuwait as the country's nineteenth province. In late July, relations between Iraq and Kuwait deteriorated rapidly, the Iraqis complaining about illegal oil wells along their joint border and massing troops in the area. By coincidence, Dante Fascell, then chair of the House Foreign Affairs Committee, had been invited to the National Geographic Society to speak to the Geographic Alliance teachers who were in Washington attending their institute. Afterward I asked Congressman Fascell about the implications of the map, the troop movements, and the oil issue. He told me not to worry. Our ambassador was on top of things, this was just a quarrel between neighbors, and the United States should not be taking sides. A few days later I happened to be in a hotel in New

York for a *Good Morning America* appearance about Puerto Rican immigration the following morning, when the phone rang at two in the morning. It was my producer. "Forget about Puerto Rico," she said. "We've got a car on its way to pick you up. The Iraqis are invading Kuwait. Ever heard of Bubayan Island? Rumaylah Oil Field? We need maps. Start writing the script in the car on the way over here." By the time we sat in the conference room, I had a rough map of the border area drawn on a yellow pad, but I wished I had that map showing Kuwait as Iraq's nineteenth province. When governments start issuing maps that include the territory of neighbors, get ready. They're committing cartographic aggression, in part to see whether anyone is paying attention. Obviously that map had raised no eyebrows in Washington. This may have been mistaken for a green light. More than twenty years later, the ripple effects continue.

Cartographic aggression takes several forms, some overt, as in the case of Iraq, others more subtle. In 1993 I received a book titled *Physical Geography of China*, written by Zhao Sonqiao, published in 1986 in Beijing. On the frontispiece is a map of China. But that map, to the trained eye, looks a bit strange. Why? Because in the south, it takes from India virtually all of the Indian State of Arunachal Pradesh, plus a piece of the State of Assam. Now this book in not a political geography of China, nor is the matter of appropriated Indian territory ever discussed in it. China's border is simply assumed to lie deep inside India, and the mountains and valleys thus claimed are discussed as though they are routinely a part of China. Make no mistake: such a map could not, in the 1980s at least, have been published without official approval. It should put not just India but the whole international community on notice of a latent trouble spot.

And so it turned out to be. Not only has China continued to insist that it is not bound by the treaty that created the boundary framework that makes all of Arunachal Pradesh an integral part of India; China has also laid more specific claim to the northwestern corner of this Indian State, the area of Tawang, linked by a mountain road from India into southern Xizang (Tibet). The basis for this claim: because the people in this corner of Arunachal Pradesh are Buddhists, and Buddhist Tibet is under Chinese control, Beijing argues that they fall under Chinese jurisdiction.

TRACKING THE MAPS OF AGGRESSION

Have a look at the entire stretch of boundary between China and India in any major world atlas, and you will see trouble from end to end. In the east, China's version of the Arunachal Pradesh issue is now printed with a note "Claimed by China." China's line, however, ends at the eastern border of Bhutan, raising the question as to its disposition westward. In the west, China controls three areas on the border of Kashmir, currently claimed by India.

This is only one part of the world where official maps convey adversarial intent, but the global importance of the neighbors involved here escalates their implications. Keeping track of what states are publishing worldwide is a costly business, but early warnings help mobilize efforts at mediation and save money in the long run. In late 1996 the *New York Times* columnist Thomas Friedman wrote a column under the headline "Your Mission, Should You Accept It" that surveyed the impact of another round of cuts to the foreign affairs budget, arguing that such cuts damaged American interests abroad and eroded U.S. global leadership. A relatively small diversion of military appropriations to foreign affairs purposes, Friedman implied, could yield huge dividends.

Friedman's column reminded me of the results of earlier budget cuts, ones that directly affected geographers—and national security. When I was teaching at Michigan State in the 1960s, one of the foreign-service positions my students could apply for was that of geographic attaché, an embassy or consular assignment whose responsibilities included the monitoring and acquisition of all maps, published officially and unofficially, in the country of record. Their tasks included analysis of the maps they acquired and assessments of their significance. "Such maps," I wrote in a lengthy response the *New York Times* published, along with a drawing by Macieck Albrecht, on October 28, 1996, "provide insights into internal problems and external intentions, the latter often an early warning of aggression." But the position of geographic attaché was eliminated during earlier rounds of budget cuts, and what Friedman was complaining about was only the latest cycle of closure of embassies, consulates, and United States Information Agency libraries. So the question arises: even as countries the world over continue to publish maps that should alert our diplomatic and intelligence agencies to risks to peace and stability, who is now monitoring them? Apparently the lesson of August 1990 went unheeded.

Keeping track of the maps being published by governments of countries or, for that matter, by the governments of States, provinces, or other sub-units (I have seen some tourist maps from regional governments that clearly make political points) is only one dimension of a larger task. The general public tends to equate "intelligence" with spying, but much critical information gathered by intelligence agencies comes from sources as mundane as local newspapers, magazines, pamphlets, transcripts, and other nonsecret publications. The problem is that these publications appear in languages and scripts requiring translation and interpretation, and all indications are that the number of foreign-language experts available is dwindling quite rapidly. Fluency in one foreign language and reading ability in another was a standard requirement, rigorously tested, when I was a geography graduate student in the 1950s. By the late 1960s, this had been reduced to one language plus ability in quantitative methods of data analysis. By the late 1970s, the language requirement had been dropped entirely in most graduate schools. Meanwhile, the number of graduate students heading for field research in foreign areas was dwindling as well. It became possible to write doctoral dissertations on, say, internal migration in India or agricultural policy in Japan from generally available information (census data or other government reports, for example) without ever setting foot in India or Japan or being able to converse in Hindi or Japanese. Geography was not alone in this trend, and the cumulative effect was to reduce the level of interaction and familiarity between young American scholars and non-Western colleagues and cultures. There is nothing like living for a substantial time in Mombasa or Chennai or Quito or Tunis, navigating the markets and bazaars, plying the bookshops and reading the local press, interacting with locals while gathering not only data but also what used to be called "field experience." When a government signals its priorities by cutting back on language-capable embassy staffs and cultural-information programs, the effects go far beyond a loss of cartographic inventory.

FACING CHALLENGES

Can geospatial technology form the needed antidote to Americans' still-endemic geographic illiteracy? Probably not, because the practical side of it will remain mostly confined to domestic use. Getting navigation help to drive to a restaurant or following a storm via satellite on the Weather Channel will not significantly broaden horizons or mitigate isolationism. Much was made, during the quick overthrow of Saddam by United States armed forces in Iraq in 2003, of the role of GIS in the conduct of the campaign, from instant regional and urban cartographics to supply-line logistics. But GIS wasn't of much help when it came to the occupation phase. The geographical concept of a "Sunni Triangle" did not appear on anybody's computer screen until the tough realities of post-Saddam Iraq set in. Those United States government officials who predicted that the roadside public would greet American forces with flowers and gratitude should have unplugged their computers and spent some time in the field.

In this second decade of the twenty-first century, the United States, still the world's sole superpower, faces challenges and threats near and far. Several of these challenges are immediate or near term; others are in the wings and will emerge later. All will be met more effectively if the American public accepts a responsibility that comes with world leadership: to be better informed about the planet geographically. An informed public, able to express its views to its representatives as Americans in this democracy are, must come to play a stronger role in the affairs of state, especially foreign affairs. After allowing Robert McNamara to steer the country into the morass of Vietnam on the basis of a perception that he knew more than we did, many Americans believed that such a thing could not happen again, that the lesson of Vietnam had been learned. McNamara may have conceded more than most Americans about Indochina, but he did not know enough. When it came to George W. Bush's intervention in Iraq, the American public was led to believe that intelligence had uncovered incontrovertible evidence of dangers that required armed intrusion. But did the public know any more about the geography of Iraq in 2002 than it did about Indochina in 1962?

In geographic perspective, the United States and its allies today confront a daunting array of challenges in a fast-changing world. Some of these challenges are beyond this (or any other) power's control or amelioration, most ominously the continued expansion of global population and the worsening social inequalities this entails. Likewise, the repercussions of planetary climate change may be alleviated, but cannot be eliminated. Even significant mitigation will fail without global leadership and international cooperation, not now forthcoming and not in the offing. Other challenges can be confronted, even by a superpower of declining global dominance, through informed policymaking and efficient use of less plentiful resources. It is in the interest of both China and the United States to prevent the development of preconditions for what would be the world's first intercultural cold war, a contest that would carry incalculable risk.

But then there are the imponderables. In a world infused by religious fanaticism and galvanized by perceived injustice, the diffusion of nuclear weapons technology is a ticket to catastrophe. The geography of terrorism is the story of spatially expanding violence, but the post-9/11 escalation foreseen by many analysts has not come to pass. Terrorist acts may cause shock and dismay, but they do not threaten global stability. The ultimate hazard, however, remains, potentially in the form of state terrorism, threatened by a nuclear-armed Iran and intimated by an intermittently belligerent North Korea, or, more ominously, in the acquisition of a nuclear weapon by an avowed terrorist organization.

If it were not for these crucial challenges, the other key events marking our transitory world would have center stage: fractious Europe's halting path toward economic integration and political coordination; petro-ruble Russia's growing assertiveness as the new Putin era dawns; fast-rising India's expanding role on its regional as well as world stage; and Brazil's halting emergence as a global economic power. It will take unprecedented leadership and uncommon public awareness to accommodate so many currents of sometimes conflicting change. And it will require a clear geographic perspective on the part of all concerned.

3

— — — —

GEOGRAPHY AND DEMOGRAPHY

THE YEAR 2011 WILL BE REMEMBERED FOR A NUMBER OF REASONS: the devastating earthquakes and tsunamis that struck Japan, the global economy in recession, the ten-year anniversary of the 9/11 terrorist attacks on New York and Washington. But long after these milestones have faded with time, 2011 will remain on the world's radar screens for two demographic reasons. First, it was some time during that year when the planet's growing human population surpassed the seven billion mark. And second, the world's urban inhabitants, for the first time in thousands of years of human history, began to outnumber those still living in the countryside.

Of course, no one knows the exact dates when these thresholds were crossed. Some calculations even suggest that the 50 percent urban ratio may have been reached late in 2010. No matter: the opening of the century's second decade should impel us to contemplate what the future will have in store. The good news is that the global population's annual rate of increase is slowing down. The bad news is that this slowdown isn't happening fast enough to safeguard the Earth from having to accommodate another three billion people before the year 2100. The good news is that the quality of life in dozens of the world's largest cities has improved markedly. The bad news is that life in many more burgeoning cities remains harsh, violent and embittering for millions of residents.

If ever the geographic question "where?" has resonance, it is in this realm of population. The demography of our planet is a mosaic of incredible contrasts and contradictions. Population data are reported to the United Nations and other agencies by the governments of the world's nearly 200 countries, but amazing social contrasts prevail *within* many of these countries, so that virtually every reported average conceals geographic differences—between regions, provinces, or other subunits—that make almost any generalization risky. In India, for example, the population of the western State of Rajasthan (70 million) is growing three times as fast as that of the southern State of Kerala (35 million). Overall, India's northern States are growing about twice as fast as its southern ones. So when India reports an "average" growth rate for its population, that average hides a wide range of social conditions. Even smaller countries exhibit sharp contrasts. South Africa, with a population only 4 percent the size of India's, is one of the most unequal states on the planet in almost every respect: growth rates among its ethnic groups, life expectancies, health conditions, incomes, housing, education. When agencies rank countries on the basis of their reported averages, social or economic, always look for the geography behind the data.

THE GLOBAL SPIRAL

The ultimate demographic generalization, of course, is based on the planet as a whole, and no matter what we have learned about this, it's an astonishing picture. A graph of world population increase makes that familiar global-warming curve look flat. After tens of thousands of years of net growth marked by surges and reversals, the human population of planet Earth did not number 1 billion until about 1820, less than two centuries ago. It took more than a century to add another billion, but consider this: in that century, after 1820 and before 1930, the human population increased by as much as it had reached in its entire history up to 1820. That alone would justify talk of a "population explosion," but it was just the beginning. The next billion accrued between about 1930 and shortly before 1960, and by the time the 4-billion mark was passed in 1975, little more than fifteen years later, alarm bells were going off worldwide. Consider this: the number of years to add one billion people to our small planet had gone from over 100 to just fifteen in less than one and a half centuries. It certainly seemed to justify all the dire warnings of worldwide famines, food wars, "standing room only" and other crises that would result.

But then something unanticipated happened. First, the food crises remained more localized than forecast, mainly because of spectacular research under the rubric of the "Green Revolution" that increased per-hectare grain yields and improved the production of other crops. Second, the population explosion—for a complex combination of reasons of which economic development, urbanization, and globalization are part—began to lose steam. It was nothing dramatic, but look at it this way: the time it now takes to add another billion people to the global population is lengthening. Things are still pretty bad: we went from 6 billion in 2000 to 7 billion in 2011, and we're likely to reach 8 billion before 2030. Specialists who study this phenomenon point out that the population base producing this increase is now much larger than it was in the 1970s, but that the rate of annual natural increase is dropping every year.

Any population's natural increase or decrease (not all populations grow all the time) is measured, obviously, by subtracting the number of deaths from that of births over a given period of time, usually reported annually. When the number of deaths per 1,000 in the population is just slightly smaller than the number of births, that population is growing very slowly (if deaths exceed births, the result is "negative population growth," the technical term for a decrease). When the number of births stays more or less stable but the number of deaths declines, the widening gap signals accelerating growth. In the world as a whole, that's the story starting in the eighteenth century, when death rates began to decline while birth rates stayed comparatively high. The widening gap was early warning of the coming population explosion, and it took a long time for birth rates to start declining as well. By the time they did, the population explosion was in full swing. In several major areas of the world (Africa, South Asia) it still has not abated.

Death rates declined because of major progress in hygiene and medicine associated with the Industrial Revolution and because of the dissemination of associated innovations from the source (mainly Europe) to the rest of the world. Just two inventions, effective soap and the toilet, contributed enormously to this; ever-better medicines had great impact as well. Refrigeration, water purification, and other advances also helped lower the death rate. Such progress ensured the survival of countless babies who would otherwise have died at or soon after birth. The death rate of a population reflects not only those who die after a substantial lifetime, but also those who die at birth or in infancy. Today, one of the most telling statistics reflecting a country's overall condition is its infant mortality rate, the number of babies that die within the first year of life, usually reported per 1,000. In 2011, some countries reported all-time lows of 3 (Japan, Sweden) but 17 African countries' infant mortality rates still exceeded 100.

Another way of looking at population growth, past and future, focuses on women's fertility rates. This may seem misleading, as it is less a biological than a social measure: the fertility rate is not a measure of the number of children a woman *could* have, but rather the number she *will* have in particular population (a region or country or province) under prevailing social circumstances. And in most of the world, the fertility rate is falling, in some countries amazingly fast. The statistical number of 2.1 is significant here: this is the so-called replacement rate, the level at which, when reached on this downward curve, population stops growing.

Between 1950 and 1970, the world's average fertility rate was about 4.5 and showed little sign of dropping significantly. But during the 1970s a downward spiral began that, by 2012, had almost halved that figure (the world average today is about 2.44). And you cannot attribute this to worldwide factors: demographers often point to Bangladesh, where neither massive urbanization nor rapid economic growth could be responsible. Nevertheless, the fertility rate in Bangladesh today is 2.16, less than half of what it was just one generation ago. And Iran is even more remarkable: its fertility rate was 7.1 in 1981 and 1.8 in 2011. Although Iran's population is still growing, this portends a future stabilizing of the population and an eventual decline.

Fertility rates, therefore, are a harbinger of the future even if current birth and death rates do not yet reveal it. Take the case of Brazil, where the fertility rate has reportedly fallen below the replacement rate. In 2011, Brazil reported a birth rate of 17 and a death rate of 6, its population still growing—at below the world average, but growing nevertheless. But Brazil's declining fertility rate points to a stabilizing population and eventually a shrinking one.

When fertility rates fall, a country's "population pyramid" changes dramatically. You can see the pyramid change over decades: fewer children mean proportionately more adults, creating a "bulge" that moves upwards and eventually reaches the top, where more and more older people survive beyond 70. But it is important to view these changes in context of the level of economic development of the countries affected. European states were already (comparatively) wealthy when this sequence of events occurred. As their populations "matured" and then "aged," the burden of the growing number of oldsters was a challenge, but countries such as Germany and France could handle it. But countries

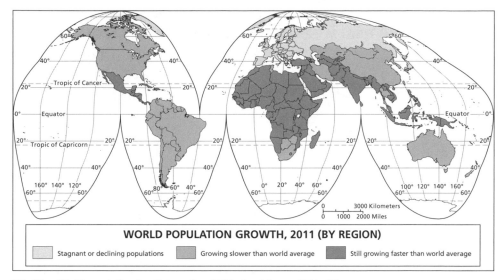

WORLD POPULATION GROWTH, 2011 (BY REGION)

Stagnant or declining populations   Growing slower than world average   Still growing faster than world average

Fig. 3-1.   World population growth by region, 2011. For data, see Table 3-1.

whose populations grow old *before* they get rich are in a very different position. For all its economic growth, the per-capita GNI of China in 2010 was little more than $6,000—less than Albania, half of Bulgaria, and one-sixth of the United Kingdom. How will China pay its state pensions and other bills? And China is a lot better off than Bangladesh or Pakistan!

NATURAL INCREASE IN REGIONAL PERSPECTIVE

Enormous differences, therefore, still exist in the well-being of this planet's regional populations, as reflected not only by infant mortality statistics but by other data as well. In the mid-twentieth century, when Paul Ehrlich and others raised the "population explosion" alarm and forecast vast famines afflicting billions across the planet, the global nature of the coming crisis tended to obscure regional contrasts, although these were identifiable even then. Today, the regional geography of population change is a key concern.

At issue is the geographically variable rate of natural increase (or decrease). Populations also grow and decline through net immigration or emigration, but natural increase is what matters here. A map of the world showing the countries where population is either stagnant (0.0 percent growth rate) or declining shows a large swath of states from Japan to Sweden in this condition (Fig. 3-1). A half century ago a movement

TABLE 3-1.   RATES OF ANNUAL POPULATION CHANGE BY GEOGRAPHIC REALM (2012) (WORLD AVERAGE = 1.2)

Subsaharan Africa	2.6
Pacific	2.0
North Africa/Southwest Asia	1.9
South Asia	1.7
Middle America	1.6
Southeast Asia	1.3
South America	1.2
Australia/New Zealand	0.8
North America	0.6
East Asia	0.5
Europe	0.0
Russia	−0.2

called ZPG (Zero Population Growth) called for this then-unimaginable goal as a global objective. It is now here, in regional dimensions, and it is fraught with its own problems. Countries with stagnant or declining populations face economic problems the ZPG crowd didn't think of.

Table 3-1 shows that South America currently is growing at the world average, with the population of Middle America increasing substantially faster and North America's significantly slower despite the massive northward migration that has become an issue in the United States. Perhaps the most remarkable aspect of the data in this table is the position of populous East Asia, now growing slower than North America. Just one generation ago there was little prospect that East Asia's rate of increase, long driven by China's high rate of growth, would plummet to this extent. But the imposition of China's one-child-only regulation had dramatic effect, to the point that China's rulers are now debating the policy's merits in the context of projections that the country's population will begin to decline. No such concerns as yet affect India, the dominant factor in South Asia's fourth-place growth rate. The same projections anticipate that India will soon become the world's most populous country.

Once again, the averages conceal some interesting details. In Subsaharan Africa, no fewer than nine countries are currently growing at 3.0 percent or more, and three of these—Niger, Burkina Faso and Uganda—

report a rate of natural increase of 3.4 percent or more, not only the highest in the world but nearly *three times* the world average. South Africa, on the other hand, grows much slower than the world rate at 0.9 percent. In the geographic realm dominated by Islamic countries, Yemen, Iraq, and Saudi Arabia are among those growing most rapidly while Turkey and Iran grow substantially slower than the regional average. In this Muslim realm, all but one small country (Qatar) grow at 1.2 percent or more; compare this to stagnant Europe, where only one country grows faster than 1.2 percent—Muslim Kosovo.

To what extent does religion play a role in these numbers? Certainly the Roman Catholic Church remains strong in Middle America (stronger than it is in South America, where other denominations are making inroads). In the past, the Vatican has sought common cause at international gatherings with Muslim clerics, seeking to constrain family-planning practices. But it is telling that population decline is most evident in the very heartland of Catholicism, even in Italy itself, where population growth in 2012 was 0.0 percent. On the other hand, the two most strongly Catholic countries in the Southeast Asian realm—the populous Philippines and tiny East Timor—have that realm's highest growth rates by far, at 2.1 and 3.1 percent respectively.

Government policies and strictures have also had mixed effect. In communist China, an authoritarian government had the power to impose its one-child-per-couple policy and to make it stick, sometimes with draconian measures. In democratic India, on the other hand, coercion in the form of government-sponsored mass sterilization campaigns led to massive resistance and voter retaliation, and the effort was soon abandoned. Today, India is growing at a rate three times as fast as China.

DILEMMAS OF DECLINE

Given the explosive growth of global population over the past century, it's no surprise that geographers and others who study demographic trends have focused their concerns on overpopulation. The prospect of billions more on a planet already strained by excessive numbers is frightening. In this context, those regions and countries where populations have become stable and are even declining would seem to offer hope for the future.

But it is not so simple. Population has tended to stabilize in countries where economic development, urbanization, and overall modernization have reached a level that puts them in the most "advanced" category (Russia, as we note later, is an exception; its population decline has other causes). And where that has happened, things have not looked so rosy. Japan's population of 127 million, for example, has stabilized and, according to demographic projections, has started a slow decline that will speed up over the next 25 years. Japan's economy, once the second-largest in the world, began to sputter. Its political system turned inflexible and sclerotic. Unprecedented labor problems eroded the fabric of society; workers who had expected life-long employment and generous benefits from their companies found themselves cast adrift when those companies could not afford to continue this. Stagnant (and thus aging) population did not cause all these problems, but the situation certainly contributed to them. As the economy slowed and government tax receipts shrank, old-age benefits had to be reduced. Economists pointed out that Japan's problems were in part self-inflicted—Japan is as tough as any country in the world in limiting immigration, for example—but there could be no doubt about it: advanced economies with stagnant populations face serious dilemmas.

Certainly the Europeans took notice. Various agencies issue predictions of population change, and while these tend to differ, they agree on one point: whatever the numbers, Europe's population (as a whole) had stopped growing. A few years ago a UN study stated that the then 27 countries of the European Union (EU) would see their combined populations decline from 482 million to 454 million by 2050. Some of the member states would shrink dramatically: Germany from more than 80 million to under 70 (and, the report warned, possibly down to a mere 25 million by 2100). More recent projections are somewhat less scary, but as the *Economist* warned, "combine a shrinking population with rising life expectancy, and the economic and political consequences are alarming. In Europe there are currently 35 people of pensionable age per 100 people of working age. By 2050, on present demographic trends, there will be 75 pensioners for every 100 workers; in Spain and Italy the ratio of pensioners to workers is projected to be 1:1." Because pensions are paid out of tax revenues, taxes will have to rise sharply to fund the generous pensions Europeans are accustomed to. Workers will demand that taxes be kept down, and labor unrest will become even more endemic in Europe than it already is. Further, countries with fairly stable populations, such as the Netherlands and the United Kingdom, will resent being enmeshed in the financial problems of other EU countries, creating potential schisms in the European Union.

One answer, it would seem, is more immigration and naturalization, but this creates its own social, political, and even economic problems. Estimates suggest that immigration (which is coming mainly from Muslim North African countries and Turkey) would have to increase five- to ten-fold to compensate for the rate of natural decrease prevailing in Europe today. The associated social strains would be unmanageable. As Europe's population shrinks and its proportion of the world population declines, dreams of an economic superpower fade. The implications of Europe's demographic data are far reaching indeed.

WHAT DOES THE FUTURE HOLD?

In important ways, Europe's demographic experience represents current circumstances in the comparatively wealthy regions of the world, but the combined population of the world's richest countries amounts to a mere 15 percent of the global total. Various agencies, from the World Bank to the United Nations, divide the world into rich and poor, but no matter how that division is made, it is clear that rich countries are growing at around 0.2 percent annually and poor countries average about 1.4 percent. The populations of the least-developed areas of the world, the 50 countries with more than 700 million poorest inhabitants (10 percent of the world's people) are collectively growing by about 2.4 percent annually. It is therefore obvious that the planet's poor will not only continue to outnumber the rich, but will do so in ever-larger proportions, at least in the foreseeable future.

From time to time, the various agencies that keep track of such data issue risky long-range projections that give rise to both optimism and pessimism. These projections tend to be superseded by revisions and modifications, and most of them should be ignored. When the population spiral began to ease toward the end of the last century, demographic forecasts held that global population growth would slow down throughout the current one, adding some 3 billion people, most of them in the poorer world, before stabilizing around the year 2100 at some 10 billion. By 2050, this reasoning proposed, global population growth would be down to 0.4 percent on its way to Europe's 0.0 percent after another half century. A study published in 1995 predicted that China's population would stabilize at 1.4 billion by 2090. But in 2011 a United Nations projection held that China would be shrinking and down to 941 million in 2100.

So all this is mostly guesswork, although the 2011 study does contain some eye-popping indicators. There appears to be little doubt that India will become and remain the world's most populous country, even if the UN is wildly wrong when it reports that India will have 600 million more inhabitants than China by 2100. And who would have guessed that the third-ranking country in 2100 would be . . . Nigeria, with 730 million? Or that the fifth-ranking state, not even among the top ten in 2011, might be Tanzania at 316 million? And get this: among the 2010 top ten, only one (No. 4 United States with 478 million) is a "rich" country today.

So much for notions that the world is "flat" or "flattening": these projections may prove to be faulty, but they confirm that the poor will form an ever-larger majority in an increasingly unequal world. The UN study still suggests that the planet's population will stabilize at 10.1 billion, which may be optimistic; but it is also clear about the rise of Africa as the world's fastest-growing geographic realm, so that the largest growth will occur in the poorest of the world's regions. The impact of this increase on Africa's already-stressed natural environments will be devastating.

Another aspect of the UN's 2011 demographic projection emphasizes how much geography matters: the global estimates mask huge contrasts between regions and among individual countries. Population increase, obviously, results from the number of children a woman has during her lifetime; statistically, a population stops growing when that fertility rate gets down to 2.1, the so-called "exact replacement rate" that produces neither growth nor decline. Fertility rates have been declining virtually everywhere, but not at the same speed nor at the same steady rate—that is, some countries experience lapses during which the fertility rate stops decreasing. This leads to the kind of surge that projects Nigeria's rise to third place by the end of the century; in 2011, Nigeria's population was growing at an annual rate of 2.4 percent, double the world average. When you look at the map of Nigeria (not much larger than Texas and Oklahoma) you're likely to say "730 million? No way!"—but when it comes to demographics, never say never. Consider this: Japan has about the same area as Montana (population: well below 1 million). Japan's current population: 127 million.

If there is one geographic-demographic datum of which we can be certain, it is this: rich or poor, the majority of the world's people will reside in burgeoning cities. A study nearly a decade old but still current reports that of the 2.2 billion increase in population between 2000 and 2030, all but 100 million will crowd into cities, so that urban areas will grow at about twice the rate of global population increase (Cohen, 2003). Although today there still are countries in which rural people outnumber urban dwellers, projections (in this case much more reliable) indicate that even in the poorer regions of the world, urbanization will reach 56 percent by 2030, which is about where the richer countries were in 1950.

In terms of sheer numbers, though, the dimensions of urban areas in the poorer world will dwarf anything seen in the rich world during its heyday. Conurbations of 50 million or more will anchor regions of India and China. Even poorer countries will see urbanization rates of 75 percent and more. Urban life will be the norm for the great majority of people in the future.

In the process, many things will change worldwide: family relationships (smaller families, proportionately more children born out of wedlock), the proportions of youngsters in the population (shrinking and destined to be exceeded by oldsters), life expectancies (longer), schooling and literacy rates (higher), secular lifestyles (more common), diets (more varied). The unanswerable question is how the countries of the richer world will relate to the greater numbers and the growing economic and military power of the currently poor.

Obviously all the estimations, projections, and forecasts I have cited must be taken with more than one grain of salt, but the fact remains: barring some unanticipated upturn in fertility, the twentieth-century's population explosion may be followed, if not by a twenty-first century implosion, then by a trend toward demographic stabilization no one even half a century ago even contemplated. What this may mean for the future of the planet is similarly beyond informed conjecture.

THE GLOBAL POPULATION MAP TODAY

So how do we display the results of all the foregoing cartographically? Like many other maps used to illustrate a moment in a constantly-changing geographic panorama, it is necessary to make some decisions. We're dealing with this topic at a pretty small scale. And when it comes

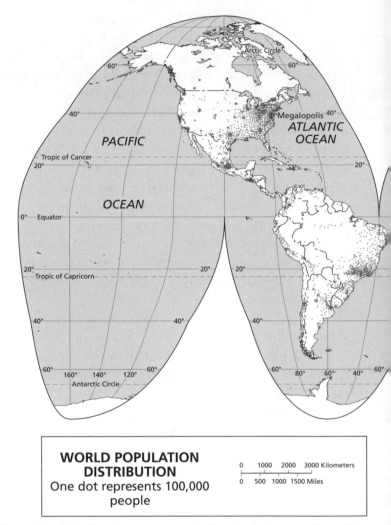

WORLD POPULATION DISTRIBUTION
One dot represents 100,000 people

0   1000   2000   3000 Kilometers

0   500   1000 1500 Miles

Fig. 3-2. World population distribution via dot map shows the three greatest concentrations, all in Eurasia, a historic pattern that endures although Africa's growth rate exceeds all others today.

to depicting the current status of world population on a map, it is well to remember that no single map can adequately represent the complexities involved. At the global scale, the best we can hope for is an instructive impression. Population can be mapped in terms of density, that is, the number of people per unit area (but for a page-size map, that unit would have to be fairly large), or in terms of general distribution, using the dot method. This is the method used in Figure 3-2, where one dot represents 100,000 people. The downside of a dot map, of course, is that the dots

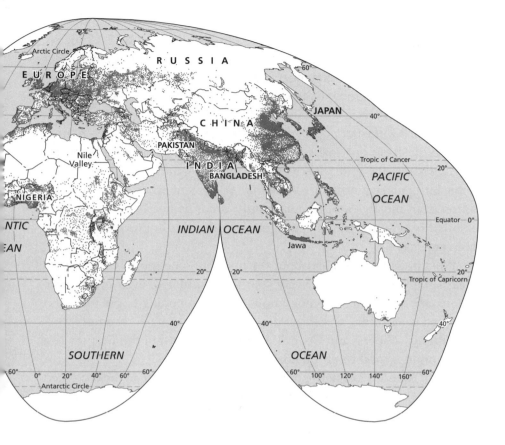

not only coalesce but overlap in certain places, for example large metro-politan areas.

Still, this map yields valuable insights. It is a powerful reminder that, on the approximately 30 percent of the planet that is land (some of it under ice), populations continue to cluster in relatively well-defined areas, reflecting the fact that of this 30 percent, two-thirds is arid, frigid, mountainous, or otherwise inhospitable to large human numbers. Of the three greatest concentrations of humanity, all of which lie on the Eur-asian landmass, two—East Asia and South Asia—lie centered on ancient river basins whose fertile soils and ample drainage supported the earli-est population explosions millennia ago. China and India are the mod-ern heirs to this ancient drama, and rivers such as the Yellow (Huang) and Yangzi in China and the Ganges (Ganga) in India are the arteries of human civilization, just as the Tigris-Euphrates and the Nile were in less-crowded Southwest Asia. In these historic regions, the majority of people still live and subsist on the land, but the inexorable migration to the cities grows and an urbanized era is in the offing. In 2012, China reported that nearly 50 percent of its people now live in cities, and India nearly 30 percent. But remember the sizes of these two giants' populations: it means that more than 650 million Chinese now live in cities and towns, and more than 350 million Indians.

The third Eurasian population cluster readily visible on the map is Europe, petering out eastward into Russia. This population, even including some still quite rural countries, is three-quarters urbanized, and some European countries are more than 90 percent urban. Here the transition from mostly rural to dominantly urban is nearly complete, and familiarity with urban life is a significant element in Europe's political convergence.

What is also remarkable about the distribution of population in Eurasia is the vast area of relatively sparsely populated, even open space still extant. Even China, synonymous with burgeoning population and crowded urban areas as well as rural areas, has a vast nearly empty interior. Russia's Siberia and Central Asia's steppes east of the Caspian Sea also remind us of the natural limits of settlement.

North America's eastern population cluster, centered on the coalesced cities of the east called Megalopolis by the geographer Jean Gottmann, is far smaller than its Eurasian counterparts, and South America is more sparsely peopled still. At the turn of the century Mexico City and São Paulo ranked, with Tokyo, among the world's largest conurbations, but urban agglomerations far larger than these are emerging in East and South Asia. Africa, too, as yet is sparsely peopled for so large a landmass—there are more people in India alone than there are all of the countries of Africa—and as the map shows, the major clusters of African population lie in West Africa, centered on Nigeria, and in East Africa, focused on the Great Lakes. The Sahara is Africa's great population void, but large parts of the landmass elsewhere remain comparatively empty as well, a situation that, as noted earlier, is destined to change. As for the continent of Australia, the mapping method on display here truly shows how sparsely peopled it remains, a function of the aridity of its "outback."

The clusters of population seen on Figure 3-2 are the repositories of the civilizations of modern humanity, the civilizations Samuel Huntington argued would find themselves in twenty-first-century conflict (Huntington, 1996). For the moment, we confine ourselves to other implications of a map that provides ammunition for those who argue that overpopulation lies at the root of most of the world's troubles.

CORE AND PERIPHERY

As noted earlier, we use map projections to enhance the utility, clarity, and relevance of maps in support of geographic discourse. Figure 3-2 is based on a much-used projection called the Goode's homolosine, devised by a distinguished professor of geography at the University of Chicago as long ago as 1923, when he used it to launch the first edition of what is still called the *Goode's World Atlas,* now in its 23rd edition (Veregin, 2010). As you can see, the advantage of this "interrupted" projection is that its large breaks lie in oceanic areas, thus reducing the inevitable distortion of the inhabited landmasses. This allows for more detail in a dot map of the kind Figure 3-2 represents. Also note that this layout of the inhabited world has Africa at the center, the Americas to the west and Australia to the east.

But there are other ways to display the whole world without breaking either land or ocean, by accepting more distortion through the "bending" of longitude lines, while keeping the latitude lines straight as the Goode's projection does. This was a popular projection in the first half of the twentieth century, when it was often used to place the Americas in the center, thus "interrupting" Eurasia and making it look as though the United States occupies the center of the world. You don't see that projection much anymore, but it has some utility in depicting the rich-poor partition of the world referred to earlier.

It is a fundamental geographic reality that human activity tends to cluster, a phenomenon of which the world's great cities and growing megacities create vivid examples. It started early in human communities and evolved into something permanent when the first towns and cities arose. When the world became partitioned into what we now call states, anchored by capitals some of which became centers of global power and influence, this had become the model of human spatial organization. In geography it is referred to as the core-periphery phenomenon, and it occurs at all levels of scale. Every country has a core (or core area), usually centered on the capital or perhaps the largest city or cluster of cities. Every country also has one or more peripheries, which may consist of productive rural areas or nearly empty land poorly connected to the core, or both. Core-periphery relationships also function in political or administrative subunits of countries such as States or provinces. If you're an American, no matter in what State you reside, you'll recognize a core (northeast Illinois, centered on greater Chicago; Massachusetts, centered on greater Boston; Georgia, anchored by greater Atlanta). Some subdivisions are large enough to have more than one core area, as in the case of California with its primary core in the Los Angeles area and its secondary core centered on San Francisco. The core area of Ontario, Canada's most populous province, is dominated by Toronto although the much smaller capital, Ottawa on the border with Quebec, also is part of it.

Core-periphery relationships not only occur at all levels of scale (even counties and parishes display them) but also exist in people's minds and show up in everyday language. Some geographic names evoke the epitome of the core: Manhattan, London, the Loop, Inside the Beltway. The same with the periphery: the boondocks, the outback, the hinterland, the wilds. More importantly, core areas tend to be richer than peripheries. Not always, to be sure: in Michigan, average annual incomes in the Detroit area today lag well behind those of the Grand Rapids area. But overall, core areas are inhabited mostly by the haves, and peripheries by the have-less.

Looking at this in global dimensions, we see the ultimate expression of those rich-poor statistics international agencies measure and report (Fig. 3-3). Using an Americas-centered projection, we can divide the world into a core area consisting of the United States and Canada, flanked by Australia and Japan to the west and Europe to the east (to get an ever more dramatic picture of this dichotomy, draw this line with a marker on a globe). Criteria for inclusion in the core include not only income but also representative government. To be sure: the global core as drawn here excludes some countries whose location and linkages affect their designation (notably Chile) and includes others by virtue of regional association (Albania), so it reflects a debatable level of generalization. But it has the merit of bringing into vivid relief the severe and dangerous inequalities marking our world.

Consider this: the global core area as defined in Figure 3-3 contains approximately 15 percent of the world's population. This population earns about 75 percent of all the money made in an average year.

Or this: 42 of the 44 cities with the highest standards of living lie within the border marking the core area. All of the world's poorest rural areas as well as all of the poverty-dominated cities lie in the periphery.

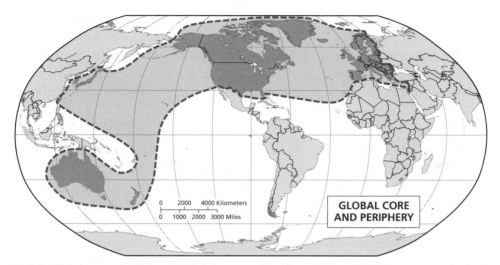

Fig. 3-3. Global core and periphery. This is only one possible representation of this phenomenon; as depicted here, the core contains about 15 percent of the world's population and earns 75 percent annually of all the money made.

And this: more than 90 percent of the 3 billion inhabitants our planet will add over the next century will be born in the penurious periphery. Millions of these will attempt to cross the core-periphery border in search of a better life.

A FLAT WORLD?

In recent years the notion that the world, if not functionally "flat," is rapidly flattening as a result of the forces of globalization has gained currency to the point of becoming a platitude. This seems to counter what is implied by Figure 3-3: so mobile, so interconnected, so integrated is this new world that historic barriers are no more, interaction is global, ever-freer trade rules the globe, the flow of ideas (and money and jobs) accelerates by the day, and choice, not constraint, is the canon of the converted. Join the "forces of flattening" and you will reap the benefits, say Thomas Friedman and others who advance this point of view (Friedman, 2005). Don't, and you will fall off the edge. The option is yours.

But is it? In truth, though the world has changed dramatically in the last 50 years, we are still parachuted by the roulette of birth into places so different that the common ground of globalization has just the thinnest of topsoil. One of some 7000 languages will become our "mother tongue"; only a small minority of us will have the good fortune of being raised in a version of English, the primary language of globalization. One of tens of thousands of religious denominations is likely to transmit the indoctrination most of us will carry for life. A combination of genetic and environmental conditions defines health prospects that still vary widely around the planet.

Some of us will be born in places of long-term peace and stability, while others will face endemic conflict in our homelands. Hundreds of millions never in their lives escape the threat of mayhem. The horizons of a life that starts in a village of a low-income tropical country differ vastly from those of an infant in a modern city of a rich country. And in every locale on this planet, even in the most favored, the combined powers of place mean something very different for women than they do for men. The rising tide of globalization may lift all boats, but most of the crews are male.

If it is obvious that the world is not flat, the question is: For whom does it appear flat? Countless world-flattening globalizers move every day from hotel lobbies to airport limos to first-class lounges to business-class seats on international airliners, laptops in hand, uploading, outsourcing, off-shoring as they travel, adjusting the air conditioning as they go. They are changing the world, these modern nomads, and they are, in many ways, improving it—depending on one's definition of progress.

But are these "globals" invariably agents of access and integration? Are they lowering the barriers to worldwide participation or raising the stakes against it? Have their influence and effect overpowered the imperatives of place, so that their very mobility symbolizes a growing irrelevance of location—and geography, in the view of more than one observer, is history?

Not yet. Even as the powers of economic globalization homogenize urban skylines from Berlin to Bangkok, another force is transforming the world, dividing it into a core of haves and a periphery of have-less or have-nots, one version of which is represented by Figure 3-3. This core includes today's three dominant "world cities" of London, New York, and Tokyo, and its economic power and political clout dominate the planet. Population growth in the global core is far below the world average; as we noted, some countries are already in demographic decline. Many countries in the periphery continue to grow at rates more than double the global average.

Small wonder that the global core is the coveted destination for millions who seek ways, legally or otherwise, to leave their abodes in the hope of finding a better future. But the core itself is taking on the worldwide exemplification of one of globalization's uglier local manifestations: the gated community. From the "security fence" between Mexico and the United States to Israel's 700 kilometers (490 miles) of walls, and from maritime patrols off northern Australia and southern Spain, and for reasons ranging from economics to safety, the global core is ringed by barricades.

Coupled with the difficulties that would-be migrants encounter when they do try to secure visas or work permits to enter globalization's fortress, these constraints are remarkably effective. United Nations data indicate that, worldwide, only 3 percent of all citizens live in a jurisdiction other than that of their birth. The overwhelming majority of the passengers of Cruise Ship Earth still die in, or very close to, the cabin in which they were born.

This means that geography and place still exert formidable power over the huge majority of the world's people, whose mobility remains constrained, their cultural baggage commonly unadaptable, their resources limited, their health imperiled, their hopes dimmed. More than a billion of these people are the poorest of the world's poor, the sickest of the sick. Another billion live on the edge of penury. At a time of reviving ideological (this time religious) extremism and dissemination of weapons of mass destruction, this is a ticket to catastrophe. Proclamations of a flat or flattening world may cheer the literati in the core, but not many beyond the barricades.

CITY AND COUNTRYSIDE: DEMOGRAPHY AND ENVIRONMENT?

What might mitigate the stresses placed on the Earth by the ongoing expansion of population in the global periphery? Some scholars argue that the accelerating concentration of population in cities will improve conditions in the countryside, reducing population pressure there and facilitating conservation efforts. Others assert that burgeoning urban centers place enormous demands on rural areas in the form of food, water, and resources, and that the pollution emanating from huge industrializing metropolitan areas far exceeds what an equivalent number of people in the countryside would generate. People who move to the cities tend to start favoring more varied diets including more meats and poultry, resulting in deforestation in the countryside to make way for pastures.

The United Nations organizes frequent conferences on environmental issues, and I remember one in particular, the 1992 UN Conference on Environment and Development in Rio de Janeiro. When I saw the agenda, I was surprised at the lack of attention given to the population issue. When I got the roster of participants (delegations consisted of political representatives, scholars, and hangers-on) I was less surprised: few if any were geographers. The population issue was not a salient topic, but it should have been. There's little point in making commitments to protect what remains of global natural environments without taking demographics into account.

The United States delegation had an uneasy time at this conference, and the president, George H. W. Bush, came under some severe criticism for not providing sufficiently strong leadership on environmental issues. But there was plenty of censure to go around. One reason so little was said about population policies had to do with the fact that the Roman Catholic Church was still very powerful in Brazil (that power has waned in recent years as evangelical churches have made inroads). Pope John Paul II had made a visit to Colombia a few years earlier and reiterated his opposition to artificial birth control, in effect exhorting Colombians, and tens of millions of listeners throughout the realm, to have as many children as they wished. This was not a good place or time to promote practices that would have a more salutary effect on future global environments than all the touted pollution-reduction programs combined.

As it happens, population growth in South America has slowed significantly despite papal exhortations and, as Table 3-1 shows, this realm is now growing at the global average and is poised to go below 1.2 percent. Once again geography matters: the key country is populous Brazil, whose rate of natural increase has halved, from 2.0 percent in 1991 to 1.0 percent in 2011. The past is reflected by Bolivia (2.0) and Paraguay (1.9); the future by Uruguay (0.5). Even as African populations mushroom, those of several South American countries are headed for stabilization.

In Middle America, though, the picture is less rosy. Here the key state, Mexico, still grows well above the world average (1.4 percent), and the realm as a whole increases even faster at 1.6 percent in 2011. Central

America displays the associated symptoms as underdeveloped countries, coping with high growth rates, see their hopes of improved living standards dashed and their dependence on foreign aid raised. Such conditions contribute to economic stagnation and the emigration of workers, resulting in friction between richer and poorer neighbors. In the richer world, efforts to control cross-border migration lead to political initiatives that are sometimes at odds with professed principles. The overwhelming approval of Proposition 187 in California during the 1994 elections reflected a rising anti-immigrant sentiment—in a nation forged of immigrants—that would crest within a decade. By 2010, the issue had come to the forefront among American priorities as the United States coped with "undocumented" immigrants numbering an estimated 10 to 12 million, and individual border States, notably Arizona, were seeking ways not only to stem the tide, but also to send back migrants who had crossed the U.S.-Mexican border illegally. This in turn led to allegations of police harassment based on personal appearance rather than actual wrongdoing. Meanwhile, various efforts to fortify the border took their own toll on America's "open" society.

As we will see, high rates of population growth constitute only one reason behind problems of economic development in poorer countries: bad government and ineffective administration, corruption, international trade rules, and other factors also contribute. And the effort to stem the cross-border migration flow between Mexico and the United States is only one manifestation of a global problem. But when populations grow rapidly, development goals and other social targets tend to stay out of reach. Maps of population growth and endemic poverty leave no doubt.

Even if human population stabilizes some time in the second half of this century, and even if it then commences an overall decline, it will be too late for much of what remains today of forests and wildlife that sustain environments and link us with our past. Biologists estimate that there may be as many as 80 million types of organisms on Earth, perhaps even more; most have yet to be identified, classified, or studied. *Homo sapiens* is only one of these, and in ten millennia our species has developed a complex culture that is transmitted from generation to generation by learning and is to some degree encoded in our genes. We are not unique in possessing a culture: gorillas, chimpanzees, and dolphins have cultures, too. But ours is the only species with a vast and complex array of artifacts, technologies, laws, and belief systems.

No species, not even the powerful dinosaurs of epochs past, has ever affected earthly environments as strongly as humans do today. The dinosaurs, and many other species, were extinguished by an asteroid impact. Some biogeographers see an analogy and suggest that the next great extinction may be in the offing, caused not by asteroids but by humans, whose numbers and demands are destroying millions of species—and with them the inherited biodiversity of this planet.

This destructiveness is not just a matter of modern technology and its capacity to do unprecedented damage, whether by wartime forest defoliation, peacetime oil spills, or other means. Human destructiveness manifested itself very early, when fires were set to kill whole herds of reindeer and bison, and entire species of large mammals were hunted to extinction by surprisingly few humans. The Maori, who arrived in New Zealand not much more than 1,000 years ago, inflicted massive destruction on the native species of animals and plants in their island habitat, long before modern technology developed more efficient means of extinction. Elsewhere in the Pacific realm, Polynesians reduced the forest cover to brush and, with their penchant for wearing feather robes, had exterminated more than 80 percent of the regional bird species by the time the first Europeans arrived. The Europeans proceeded to ravage species ranging from snakes to leopards. Traditional as well as modern societies have had devastating impacts on their ecologies, and on the ecologies of areas into which they migrated.

Is wanton destruction of life a part of human nature, whatever a society's cultural roots? The question is as sensitive as questions about racism and sexism. Still, regional differences in attitude and behavior can be discerned. African traditional societies hunted for food or for ceremonial reasons, but not for entertainment or amusement. The notion of killing for fun and fashion was introduced by Europeans. Hindu society and religious culture in India are more protective of the natural world than many others. The extermination and near extermination of many species of animals in India took place during (Muslim) Moghul and European colonial times.

Malevolent destruction of the environment continues in various—indeed many—forms today, ranging from the deliberate setting of oil fires by Iraqis during the 1991 conflict over Kuwait, to the mercury poisoning of Amazonian streams by Brazilian gold miners and the disastrous oil spill in the Gulf of Mexico from a British Petroleum Company oil rig in 2010. For the first time in human history, however, the combined impacts of humanity's destructive and exploitative actions are threatening the entire Earth's biodiversity. Most of that biodiversity has been concentrated in, and protected by, the great equatorial and tropical rainforests of South America, Africa, and Southeast Asia. Now the onslaught on this last biogeographical frontier is under way, and for the future of the planet, the consequences may be calamitous.

MALTHUS ON THE MARGIN

When the English economist Thomas Malthus in 1798 published his warning that the population in Britain was growing faster than the means of subsistence, he predicted that population growth would be checked by hunger within 50 years, leading to the disintegration of the social order. For three decades after sounding the alarm, Malthus faced severe criticism from those who saw the future differently, but he gave as good as he got. The exchange is one of the most interesting debates ever recorded. In the end, both Malthus and his critics were proven wrong. Food production has not, as Malthus prophesied, increased in linear fashion: it has grown exponentially, recently during the dramatic Green Revolution and today through genetic engineering of an even more invasive kind. And populations did stabilize, though not because of lack of food.

Twentieth-century apostles of Malthusian credo, called neo-Malthusians, forecast scary scenarios of a twenty-first-century planet with tens of billions of inhabitants in a continuous, mortal struggle for the means of survival. The notion of "doubling times" for growing populations was posited as though it conformed with reality: since Brazil in 1970 had about 100 million inhabitants and was growing at 2.8 percent, and since a population growing at 2.8 percent would double in 26 years, Brazil would have 200 million inhabitants in 1996, 400 in 2022, and 800 around the middle of this century. The fact that the growth rates of many European populations were already declining was not deemed relevant: the European demographic "model" was not applicable to the rest of the world. Today, you don't hear much anymore about doubling times. Growth rates virtually everywhere are declining, and despite UN forecasts even Nigeria is unlikely to sustain the growth rate that is required to propel it through two doubling phases.

Nevertheless, the shadow of Malthus hangs over the global periphery, and it is not certain that widespread hunger and famine have been permanently banished. In 2011 a devastating drought in Africa's Horn and East caused a deathly famine resulting from a combination of environmental and political causes. Simultaneously, UN estimates indicated that worldwide more than 800 million people, a majority of them children, continue to suffer from malnutrition or worse. Many of these people pay the price for being in the wrong place at the wrong time, because the truth is that—barring disastrous climate change—all people on Earth could be fed. Diets might not be adequately balanced, but they would ensure survival if there were ways to distribute available food and to make it affordable. Tragically, that is still not happening. True, many millions who live in remote areas depend almost totally on what they can grow and raise, people who remain subject to the vagaries of the environment and suffer when droughts or other natural disasters strike. But millions more go hungry because governments fail to achieve stability or security. Terrible civil wars in Equatorial Africa caused widespread dislocation and resulting starvation over the past two decades; to the north, the regime ruling (then-) Sudan used food as a weapon of war, denying it to millions of refugees from the decades-long war between the Muslim North and the Christian-animist South. Even as the aftermath of that war lurched toward a solution (an independent South Sudan seceded from the North in 2011), a new humanitarian crisis, also precipitated by the regime in Khartoum, made the geographic name *Darfur* synonymous with genocide and famine.

Government failure to protect people from hunger is not an exclusively African phenomenon. Reports of widespread starvation resulting directly from government policy in North Korea have become commonplace for that blighted country. The Soviet Union's punishment of Ukrainian peasants during and after the Bolshevik victory cost millions of lives. Taliban rule in Afghanistan in the 1990s created severe malnutrition in areas opposed to their fanatical practices. But politics and ideology are not the only threats to powerless people on the global margins. Government economic policies also play a role. It is not enough to produce a quantity of food for people to sustain themselves; they must also be able to afford to buy it. Well-stocked markets most of whose local customers cannot pay for a pound of grain do not reflect an absence of malnutrition.

While the final surge of Malthusian population explosion may not reach the limits the English economist anticipated, there is some hope that a less dramatic scenario lies in store.

It is always possible that the whole process will be derailed by some catastrophic event—a global natural disaster such as an asteroid impact, abrupt climate change, a pandemic of some unstoppable disease, an outbreak of nuclear war—but if things continue as they are now, it is conceivable that the world may be planning for a ZPG environment in a half century or so. Of course this will produce its own problems. Who will do the work? How will the costs of longevity and lengthy retirements be paid? What will happen to the concept of Social Security? So far, the European model certainly is no guide. Still, it is conceivable that hunger will at last have been defeated and the gap between the rich and the poor (there will always be one) will narrow.

Even as we catch a glimpse of light at the end of the population tunnel, another concern arises. After the half century of comparative stability during the two-superpower Cold War, and despite the proxy conflicts that Cold War entailed, the world is now a far more volatile place, in which further nuclear proliferation threatens, civilizational compulsions drive new forms of conflict, and economic globalization meets cultural mobilization on a new and dangerous battleground. A slowing population spiral creates opportunities for a lessening of global inequalities. Herein lies hope for the future.

4

GEOGRAPHY AND CLIMATE CHANGE

LET'S GET RIGHT TO THE POINT. THERE IS NO SUCH THING AS "global" warming. Make no mistake: the evidence for *average* planetary temperature increases over the past several decades is incontrovertible. The Earth, on average, is significantly warmer today than it was, say, 40 years ago. But when our planet warms up, there are always places where it not only stays cool but may even get cooler. Such exceptions, when they are not understood to be part of a process that has been going on for as long as the Earth has had an atmosphere, may seem to cast doubt on theory behind the dominant trend. "They had the coldest winter in New Zealand in a century" or "this was the coolest summer ever in Tuscany" do not invalidate the premise that the Earth *as a whole* is experiencing a sustained warming trend. The bitter winter in Europe in 2012—the coldest in more than 100 years in some areas—did not portend any global climate reversal. In the U.S. Northeast, the winter was milder than most.

When the Earth cools, as it has done numerous times during its geologic history, the same is true. "Global" cooling does not mean that the planet gets colder everywhere at the same time. Even when the glaciers stage a comeback and push into lower latitudes, some places in higher latitudes stay mild and even livable. We know this because our distant ancestors migrated into Europe during a very cold time, making use of places where plants and animals survived despite the "average" cold and migrating and adapting when necessary. A newspaper headline 30,000 years ago might have read: "Bitter Cold Grips North America. Mild Conditions Continue in Central Europe."

So geography matters when it comes to climate change. Few topics have aroused as much public debate and dispute over the past quarter century as has global warming (to use the popular shorthand). It has become more than an argument over information such as temperature readings and carbon-dioxide measurements, melting mountain glaciers and thinning polar ice. The global-warming issue pits scientists against politicians, environmentalists against energy-company representatives. The public is bombarded with dire warnings about rising sealevels, disastrous hurricane seasons, devastating droughts, fearsome floods. Political leaders cast doubt on the science.

It is not surprising that many people do not know whom to believe. If large percentages of Americans cannot identify major physical or political features on a blank map, even fewer could be expected to be able to outline the reasons why atmospheric pressure systems form and move the way they can or ocean currents flow the way they do. An introductory college course in physical geography marvelously summarizes the essentials of the global-warming controversy, because it deals with all the interacting mechanisms that comprise the planetary system, from the evolution of continents to the impact of ice ages and from climate change to biogeography.

Certainly the media—newspapers, magazines, television—have contributed to the obfuscation. Even responsible and respected journals seemed susceptible to hyperbole, and objectivity became a casualty of their certitude and fervor in support of theory. Make no mistake: despite regional variations, the Earth is warming. Human activity in its massive and diverse dimensions is contributing to the current temperature trend. There is no need to exaggerate or accuse, even in the interest of swaying the views of those who may not be persuaded.

Because they know that human action is a factor in the present phase of planetary warming (the science, to be sure, is incontrovertible), some advocates of mitigative action inadvertently contribute to the antagonism that continues to mark the public debate. In the interest of persuasion, Al Gore's movie, *An Inconvenient Truth*—well-intentioned as it is—skirts the "truth" enough to give critics some solid ground. Paul Krugman, the *New York Times* columnist, should have tempered his assertion that people who "deny" global warming are treasonous and immoral (Krugman, 2009). Among those responding, a respected senior scientist wrote: "Having done scientific research for four decades about possible environmental effects of global warming, and having been one of the earliest ecologists to be concerned about such possibilities, I am shocked and affronted by Paul Krugman's column 'Betraying the Planet'" (Botkin, 2009). Of course Mr. Krugman is an economist and a television commentator, not a scientist, and while he borrowed effectively from geographic research and theory in his earlier work (Berry, 1999), he cannot be expected to understand the complexities of climate-change science, which straddles physical and human geography and ranges from computer forecasting to mitigative policy mechanisms. To get at the answers that must be found, debate is crucial and accusations are unhelpful. As Botkin writes, the effect of Krugman's article "is to suppress any open discussion of global warming."

In truth, Krugman wrote his intemperate column in frustration after witnessing the narrow passage of a climate-change bill in the 2009 Congress, where 212 Representatives voted no and many rejected the entire premise that planetary warming relates in any way to anthropogenic greenhouse-gas emissions. Indeed it may be the case that no scientific debate in American history has done more to devalue science in the public eye than this dispute over climate change. And let there be no doubt: scientists themselves have contributed to this picture—by failing to educate as well as opine and, in some instances, by seeking media celebrity and inadvertently sowing doubt. One of the issues linked to global warming involves weather extremes—the prospect of a rising incidence of hurricanes, floods, droughts, tornadoes, and other threats resulting from increased global temperatures. Some years ago a team of climatologists at Colorado State University began to forecast the quantity and intensity of hurricanes that would form over the Atlantic Ocean and the number that would strike the North American mainland. Every spring, this prediction received wide television and press coverage, with repeated allusions to enhanced global warming as a causative factor in the growing threat being faced. Over time, however, these scientific predictions, demonstrably the result of sophisticated and complicated modeling, missed their mark so widely, and so often, that they became the target of late-night comedy routines. It seemed that these forecasts would have been better concealed in the scientific literature as pioneering efforts in the difficult field of long-range forecasting rather than being given public exposure with all the associated risks.

More seriously, such miscalculations proved to have grave consequences. The initial (later revised) forecast for the 2006 hurricane season—the prospect of unusually high rainfall resulting from several hurricanes sweeping across Florida—impelled engineers responsible for the water level of Lake Okeechobee to open the floodgates and lower the lake level by some 45 centimeters (18 inches) in order to prevent serious and dike-damaging floods during the projected "active" season. When the hurricane series failed to materialize and a prolonged drought desiccated the Southeast, the *Miami Herald* in December 2007 reported that Lake Okeechobee and its ecosystem had suffered severe damage as large parts of the lake bottom lay dry, and that Florida stood to lose millions of dollars in crop losses.

None of this, though, dissuaded the Colorado State University scientists from pursuing their objectives. In April 2008 the media carried their next forecast, described by the *Wall Street Journal* as based on "improved forecasting methods after two years of dire forecasts didn't pan out." There would, according to the new model, be 15 named storms, 8 of which were to become hurricanes, and 4 of these would develop into major storms with winds of at least 178 kilometers (111 miles) an hour. This forecast resembled the ones for 2006 and 2007, and the public would be excused for assuming that chance, not science, might finally make the scientists look accurate in what the *Miami Herald* scathingly referred to as "voodoo meteorology." None of this helped build public confidence in the science behind climate change.

CYCLES AND SURGES

In truth, so many forces are at work in the climate-change arena that it is amazing that any long-range forecasts can be made at all. The warmth the Earth receives from the Sun varies according to the angle of rotation of its axis and the direction in which that axis points; the first involves a 41,000-year cycle, the second a cycle that varies from about 23,000 to 26,000 years. The Earth's orbit around the Sun is not steady but varies as well, an eccentricity that takes the Earth farther and nearer the Sun in a cycle of more than 400,000 years. The inclination of that orbit also changes, this cycle ranging from about 100,000 to more than 400,000 years. Then there is the Sun itself, about whose long-range cycles little is yet known. The 11-year "Sunspot" cycle is well understood, but it is interesting to note that a cold phase in Europe between 1645 and 1710 appears to have coincided with a period of low-Sunspot activity known, after its discoverer, as the Maunder Minimum. Then in the early 1800s there was a less-pronounced minimum (the Dalton Minimum); it is not clear whether these minima are part of a solar cycle still to be deciphered. Add to this a tantalizing history of "turnovers" in ocean circulation, and you can see how challenging any forecasting of Earthly climate inevitably is. One of the most interesting developments in recent years has been the realization that the cycles just mentioned seem to dominate sequentially, that is, the timing of warm and cold spells may depend on what cycle (or combination of cycles) has the upper hand. For example, over the past million years, it appears that 40,000- to 50,000-year swings from warm to cold and back again dominated until about 425,000 years ago, when quite suddenly the cycles changed to 100,000-year alternations. As we will see later, this has huge implications for what we are experiencing today.

Could nature overpower the human-enhanced global warming occurring today and put climate change in reverse? It would be the ultimate arrogance to assert that human action could surmount the power of nature. But that is no excuse for not doing what should be done worldwide: to limit the pollution poured into the atmosphere in this industrial era of environmental destruction. So let us put the "global-warming" issue in geographic perspective, and take a chronological and spatial look at how we got here, environmentally speaking.

But before we get started, we should—just for this chapter—get used to thinking in terms of millions and billions of years, which isn't easy. One way to go about this is to relate our planet's age to our own. If you happen to be in your mid-forties, it is easy: a year in your life represents 100 million years of Earthly history. One month equals about 8.3 million years, and a week, just under two. A single day in your life equals some 275,000 years, and one hour about 11,000. Consider this: the emergence of modern humans has taken place, comparatively, in the last day of your life; the rise of modern civilizations, during just the past hour.

If you are in your early twenties, just double these figures; if you are in or near your late sixties, subtract about one-third. No matter what one's age, though, the recency and brevity of our human ascent and domination of this planet cannot but impress. No matter what your age, the dinosaurs held sway until less than a year ago!

DRAMATIC BEGINNINGS

Some 4,600 million years ago planet Earth congealed from an orbiting band of cosmic matter into a fiery ball of molten substance burning fiercely and emitting clouds of superheated gases that found its place in the evolving solar system as the third planet among the nine revolving around the Sun. Millions of lightning bolts struck the red-hot surface while, inside the young planet, heavier matter sank toward the center and lighter material accumulated in the outer layers, all of it kept in constant motion by the intense heat.

And then, when the Earth was a mere 100 million years old, a cataclysmic event changed it forever. In the continuing chaos of the evolving solar system, a large object, perhaps as large as Mars, approached our planet on a collision course. Even a thick, protective atmosphere would not have cushioned the Earth against the devastating impact that followed. The planetoid struck at a low angle, a glancing blow that briefly buried it in the molten mass of Earth's primordial shell. So great was the speed of the object, so huge was the collision, that much of it bounced outward again into space along with a large volume of Earthly matter.

But it did not fly out into space. Slowed, weighed down, and unable to escape the Earth's gravitational field, this giant ring of matter soon coagulated into a single ball that orbited the mother planet. The Earth had acquired its Moon.

Imagine the scene, 4,500 million years ago. Just a few hundred miles above the Earth's surface hung an incandescent Moon that filled the night sky from one horizon to the other, seemingly so close to the Earth that you could touch it. A gaping craterlike depression marked the place where the impact had occurred, threatening for a time the very structure of the planet. The low-angle blow from the impact object set the Earth spinning on a wobbly axis, so fast that one rotation may have lasted only about four hours. The force of this rapid rotation set up wild currents of motion in its outer as well as inner layers.

Yet our planet held together, and the Moon's orbit grew progressively larger during the several hundred million years that followed. By about four billion years ago, the Earth's rotation had slowed significantly as well, so that our planet's day had lengthened to around ten hours, and the Moon was nearly half as far away as it is today. (The Moon continues to move away from the Earth in very small but measurable increments.) At the same time, patches of the Earth's crust began to cool enough to harden molten material into the first solid rocks. Initially, these patches soon were melted down again by streams of hot lava, but eventually some of them survived. The Earth had begun to form a crust.

Not much survives from these ancient "rafts" of solid rock. Earth's continental landmasses are continuously recycled, pushed and pulled

WEGENER'S CONTINENTAL DRIFT HYPOTHESIS

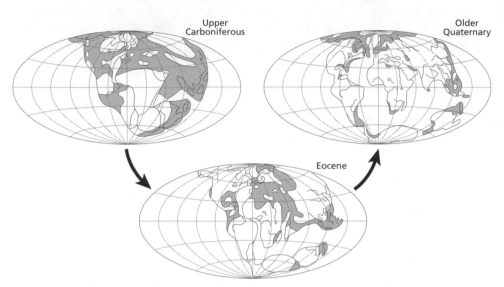

Fig. 4-1. A modified version of Wegener's original map. He had the dates wrong and the mechanism he proposed to account for the "drift" of continents was a failure, but his hypothesized supercontinent, Pangaea, did exist as he postulated it.

below the crust, heated, melted, and regurgitated along midocean ridges and trenches, so it is remarkable that rocks several billion years old do survive in a few places, including Western Australia and interior Africa. But the familiar continental outlines we see on globes and in atlases today are nothing like their antecedents three to four billion years past.

Not only are whole continents recycled: they move horizontally in the process the climatologist-geographer Alfred Wegener called continental drift. A century ago Wegener, observing the close "fit" of the shapes of continents across the Atlantic Ocean, thought that this was unlikely to be matter of chance. The landmasses had once formed part of a supercontinent, he reasoned, that fractured into the continents we see on the map today. He called this hypothetical supercontinent Pangaea, and his map of (Fig. 4-1) is one of the most prescient ever drawn (Wegener, 1915).

Wegener's hypothesis engendered the later theory of plate tectonics and crustal (sea-floor) spreading. Scientists now know that Pangaea and its breakup were only the latest episodes in a cycle of continental coalescence and splintering that spans billions of years. This latest Pangean fragmentation, however, began only about 180 million years ago and continues to this day. Where great slabs of the crust called plates collide and continental margins are pulled under, as is happening along the west coast of North America, earthquakes and volcanic eruptions accompany the process. That is why we recognize a circum-Pacific "Ring of Fire," marking these gigantic collisions from Chile to Alaska to Indonesia to New Zealand (Fig. 4-2). The earthquake that caused the December 26, 2004, tsunami (its epicenter is marked by the arrow off northwest Sumatra) was the result of such plate collision. The earthquakes off Japan's east coast on March 11, 2011, triggering tsunamis that killed some 16,000 people, had similar origins. Our planet is still young and restless.

So the continents, made of the lightest rocks (solid or molten) on Earth, ride like rafts on the mobile, heavier plates below. And what makes those slablike plates move? That was Wegener's unsolved problem: the mechanism for his continental-drift theory. The answer came from an unlikely place: the ocean floors, where upwelling, red-hot lava creates new crust even as old crust elsewhere is pushed under (the process is called subduction) as plates collide. Wegener kept comparing North America and Europe and South America and Africa, but the crucial answer to his unsolved problem lay midway between them: along the Mid-Atlantic Ridge, where the Transatlantic continents were once conjoined. There, heavy, dark, basaltic rock forces its way upward, pushing earlier, hardened lava aside. We can actually observe the process occurring today near Iceland,

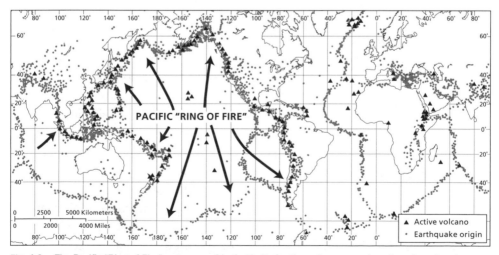

Fig. 4-2. The Pacific "Ring of Fire"—an ocean-fringing belt of active volcanoes and earthquake epicenters—marks the planet's most unstable crustal zone. Note the traces of the Mid-Atlantic Ridge and other midoceanic ridges in the Indian and South Pacific oceans, where new crust is formed.

where the normally submerged Mid-Atlantic Ridge rises to the surface. New land is being formed as islands of lava rise above sealevel.

All this is not just hypothetical any longer. We can now measure the movement of continents. One year from now, the room in which you are reading this will be about 13 millimeters (a half inch) from where it is today, assuming you are somewhere in North America. That may not seem to be much, but calculate what it means over geologic time. In just 1 million years, the distance will be 13 kilometers (8 miles). But the breakup of Pangaea started about 180 million years ago—and the North American plate is by no means the fastest-moving one. So the continental landmasses, and the plates that carry them, have moved thousands of kilometers since then.

What will the map of the distant future look like? At present the Earth exhibits the kind of pattern that attracts geographers' attention: so uneven is the distribution of landmasses that we routinely speak of a "Land Hemisphere" and a "Sea Hemisphere." This, of course, was far more pronounced when Pangaea still existed. Plate motion since then has already lessened the Land-Sea Hemisphere contrast. But the Pacific and its oceanic plates still cover nearly half the planet's surface, and the process has a long way to go. It may be, however, that plate movement is slowing down, and some geologists and others working on this problem suggest that continental motion may actually come to a halt, after which the landmasses may once again converge to form still another supercontinent. So "continental drift" may in fact be a cyclic process that has been moving the landmasses for as long as the Earth's crust has had plates and continents.

There is some evidence for this notion here in North America. As we noted, when plates converge and collide, the lighter continental plate overrides the heavier oceanic plate, but parts of both are pushed downward during subduction, and continental crust, with its fossils and telltale structures, is lost. That is happening now along North America's west coast, where spectacular coastal scenery bears witness to the forces at work. But before Pangaea broke up and North America started its westward journey, the continent was headed the other way, being carried toward the Pangaean cluster of landmasses. The coast of today's eastern United States was jammed into Morocco and other parts of West Africa, and the East, not the West, had the spectacular scenery associated with subduction. The Appalachian Mountains bear eroded witness to that pre-Pangaean time.

Over the long term, therefore, think of our planet's surface as ever changing, of continents moving and the crust shaking, of oceans and seas opening and closing, of land lost by subduction and gained by eruption. And this is only one dimension of the ceaseless transformation of Earth that began 4.6 billion years ago.

OCEANS PAST AND FUTURE

The fall of the Berlin Wall in 1989 led to much introspection—not only political, but also philosophical and scientific—and gave rise to a spate of books signaling the onset of a new era. Their titles were often misleading, such as *The End of History* by Francis Fukuyama, but none more so than one by John Horgan (1996) called *The End of Science*, which argued that all the great questions of science had been answered and that what remained, essentially, was a filling of the gaps. When it comes to global environments, however, some great questions remain open.

One of these relates to the oceans. Planet Earth today is often called the Blue Planet because more than 70 percent of its surface is covered by water and views from space are dominated by blue hues and swirls of white cloud, but in truth we do not know with any certainty how the Earth acquired its watery cloak, or exactly when. Some scientists hypothesize that the water was originally trapped inside the Earth during its formation and rose to the surface during the time when heavier constituents sank to form the core. The gases that are released during volcanic eruptions are mostly (more than 95 percent) water vapor, and massive volcanism marked Earth's early history, though lessening over time. Others calculate that most of the water that did reach the surface in this way would have been evaporated into space by the searing heat then prevailing, suggesting that another source must be identified. This has led to the comet hypothesis, which proposes that icy comets bombarded the Earth for more than a billion years while its atmosphere was still thin, accumulating fresh water from space that filled the basins in the formative crust. But studies published in late 2004 report that the chemistry of the oceans cannot be matched to that of icy comets (now much better known than before), casting doubt on the comet hypothesis.

Obviously, the "end of science" has not arrived when it comes to as crucial a question as this, and here is a related one: will the Earth retain its life-giving oceans permanently? Probes of our neighboring planet Mars produced some startling conclusions: Mars may have lost a global ocean averaging more than 30 meters (100 feet) deep, and there are indications that Mars at one time had even more water (as a proportion of mass) than planet Earth. Why and how rapidly did Mars lose its ocean? And what may that loss portend for Earth?

ICE ON THE GLOBE

I always suggest to my students that they should be as familiar with the geologic time scale as they are with the months of the year and the days of the week—it is a great way to keep things geographic in temporal perspective. (Table 4-1). Geologists refer to the first 800 million years as the Hadean, and indeed the Earth was hot as Hades during that eon although recent research suggests that it cooled faster than had been thought. The next 1,300 million years form the Archean, when the oldest surviving continental rocks and the first life forms are recorded. Next comes the Proterozoic eon, lasting from 2,500 million until 570 million years ago. It is late during this eon that something dramatic appears to have happened: the Earth went into a deep freeze.

The hypothesis is known as Snowball Earth, and the evidence is coming from rocks as far apart as China and Australia. It suggests that the Earth did not just cool, as has happened several times since: apparently the entire planet froze, from pole to pole and from land to sea. The landmasses lay buried under ice and snow; the ocean surface was frozen solid. What might have caused this to happen? A temporary but significant decline in the Sun's radiative output is one possibility. A rapid decline in methane-producing microorganisms (methane was the key early greenhouse gas) with the rise of oxygen-generating microbes around 2,300 million years ago may have chilled the entire planet (Kasting, 2004). Some scientists suggest that the stabilizing crust and consequent decline of volcanic activity could be a factor. Whatever the cause, the planet and its early life forms experienced a crisis. Whether the Earth was indeed a "snowball" or, as others suggest, a less frigid "slush ball," the days of sustained warmth as its hallmark were clearly over.

Whatever the outcome of the search for evidence relating to the Snowball Earth theory, we know that this was not the last time the Earth experienced an ice age. Several have followed, and the most recent one is in progress right now. All known ice ages, and perhaps even the Proterozoic one, have periods of severe cold separated by shorter phases of comparative warmth. We are experiencing such a warm interruption at

TABLE 4-1.   THE GEOLOGIC TIME SCALE

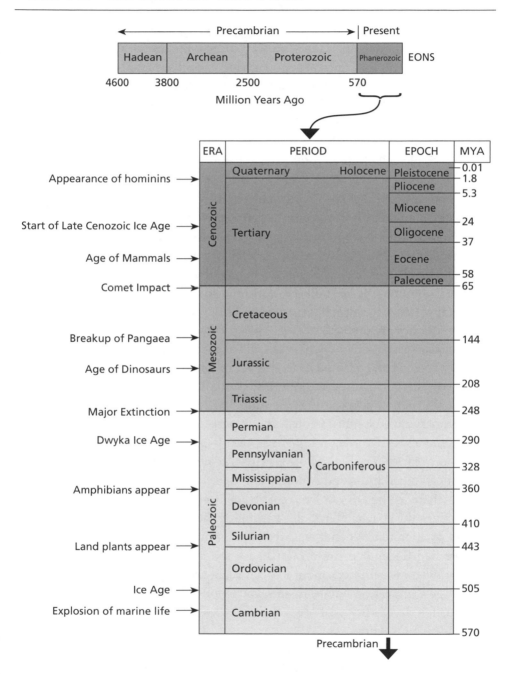

this time, one that has lasted roughly 12,000 years—roughly, because the rapid warming that brought us today's mild climates actually started about 18,000 years ago but was sharply interrupted about 12,000 years ago, when frigid temperatures briefly returned. But the ice age of our time began about 40 million years ago and, as we will see later, continues.

What we do know about ice ages is that their often rapid environmental swings pose challenges to all life forms, from eukaryotes to humans. Ice ages are times of accelerated evolution; organisms that adapt tend to survive, those that cannot, perish. During the Snowball Earth Ice Age, single-celled eukaryotes evolved into multicelled, more complex organisms. Those with protective shells (there was plenty of calcium carbonate around) had a better chance of survival. When the ice age ended, life on Earth was set for its Cambrian Explosion, the burgeoning of marine organisms in unprecedented diversity that marks the oldest period of the Paleozoic era. Then, after more than 4,000 million years of the planet's existence, began the drama that would lead to the emergence of humanity, 570 million years later, again during an ice age.

How many ice ages have occurred between the Proterozoic and the current one? We are sure of one at least: the ice age that occurred when Pangaea as a supercontinent was still in one piece (Fig. 4-3), and it may have been unprecedented in its impact on life on Earth. This happened during the Permian period, the last of the Paleozoic era, between 290 and 251 million years ago. At the time, Pangaea consisted of a northern periphery called Laurasia, consisting of parts of Eurasia and North America, and a southern sector known as Gondwana, at whose core lay what is today Africa. The South Pole was just offshore from South Africa; Australia adjoined India and Antarctica. When the Permian Ice Age (known elsewhere as the Dwyka Ice Age) struck, forests were widespread, amphibians thrived, small reptiles had made their appearance, and insects had proliferated. When it was over, one of the greatest mass extinctions ever had decimated life on Earth. As the map shows, a vast area of Gondwana was buried under ice, but what the map cannot show is the ice age's impact on environments far beyond the high latitudes. Huge areas fell dry, forests withered, and countless species of plants and animals became extinct.

That, at least, was the state of knowledge until 2003, when geologists produced evidence indicating that the Permian period ended in catastrophe, not frigidity. Asish Basu and a team of researchers found telltale bits of meteorite in Antarctica along with other indications that a huge

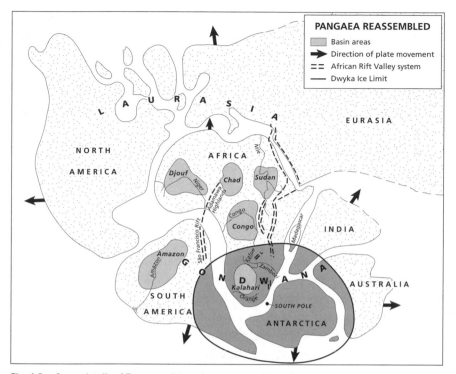

Fig. 4-3. Some details of Pangaea, the ancient supercontinent, and its southern sector, called Gondwana. The circular area in the south is the glaciated part of Gondwana; the South Pole lay between present-day Antarctica and South Africa.

meteor struck the Earth about 251 million years ago, killing as much as 90 percent of all life on the planet (Kerr, 2003). Rather than succumbing to ice-age cold alone, animals as well as plants were incinerated by the impact and its aftermath, which may have included huge volcanic eruptions pouring out vast volumes of molten rock from fissures and vents in the shaken crust. The scientists hypothesize that this may have been the most devastating of the Earth's five known mass extinctions, resulting from a combination of environmental, extraterrestrial, and geological forces that ended not only the Permian period but also the Paleozoic era (Table 4-1).

When that ice age ended and the Mesozoic era opened, there was little left of Permian life. But now the post-ice-age planet made up for that. Tropical warmth replaced Arctic cold, moisture and precipitation abounded, atmospheric oxygen increased markedly as luxuriant forests spread and the Earth was ready for the faunal exuberance of Jurassic Park. The age of the dinosaurs also saw the first birds, the first marsupials (animals whose females nurture their offspring externally in a pouch, not internally in a placenta), and the first angiosperms (plants whose seeds are encased in fruit).

Even the massive, sky-blackening volcanism that accompanied the breakup of Pangaea during the Jurassic failed to spoil the party. The dinosaurs grew larger and larger, specializing into herbivores and carnivores and competing fiercely for survival. As the landmasses separated and the seas between them widened, species found themselves isolated and evolved into distinctive forms. Only another ice age, it seemed, could end the Mesozoic's profusion.

SUDDEN DEATH

The diversity of dinosaurs and the flourishing of Mesozoic plants reached their zenith during the Cretaceous period, when huge birds flew in vast forests and flowering plants spread across the world. It was warmer even than it is today; polar and mountaintop ice were long gone and dinosaur species roamed from Alaska to Antarctica. Small mammals managed to survive in special niches, but the day belonged to the giant reptiles whose only enemy, it seemed, would be a sudden return to the frigid conditions of the Permian.

But the age of the dinosaurs came to a much more dramatic end— not with a glacial whimper but with an extraterrestrial bang. One day about 65 million years ago, a comet or asteroid only about 10 kilometers (6 miles) in diameter streaked toward Earth at a speed of 90,000 kilometers per hour (55,000 mph) on a collision course. It approached from the southeast at a low angle, striking Earth in what is today the area of the Yucatan Peninsula of Mexico. When you get off a boat at the small port of Progreso, there is a small, hand-painted sign that points to Chicxulub, a Maya name for a local village. But to geographers, Chicxulub means the end of one era and the start of another. Here the asteroid's impact produced an explosion equivalent to about 100 trillion tons of dynamite, forming a crater approximately 180 kilometers (110 miles) in diameter, 65 kilometers (40 miles) deep, and encircled by a geological fault 30 kilometers (about 20 miles) beyond, all of it buried today by later sediments.

It is possible that the Chicxulub asteroid was one of a swarm, and that smaller ones struck the Earth elsewhere, including the ocean. In any case, the impact's devastation reached around the planet, and was at its worst in North America. The impact area was a shallow sea with soft, deep sediments, and the blast sent a mass of debris hurtling thousands of miles into the heart of the continent and high into the atmosphere and beyond. Researchers David King and Daniel Durda calculate that some of it reached halfway to the Moon before falling back to Earth. And when it did fall back, it rained red-hot rocks on the rotating planet, setting fires to forests almost everywhere. The atmosphere was heated enough to evaporate entire lakes, incinerate whole ecosystems, and extinguish most life over large low-latitude regions.

The Chicxulub impact ended the Cretaceous and marked the beginning of a new geologic-calendar period, the Tertiary. Popularly, the transition is called the K/T Boundary, but its significance is hard to overstate, because this was one of the three greatest known mass extinctions ever. While it is possible that some dinosaurs survived the original blast, notably in higher latitudes, food chains had been fatally disrupted and they, too, died out. Some small mammals were better equipped to outlive the crisis, perhaps keeping cool in high-latitude caves and burrows, depending less on the luxuriant vegetation and reptilian life the dinosaurs had needed. But the faunal and floral exuberance of the Mesozoic era came to a sudden, irrevocable end.

The K/T blast had long-term effects on global environments. Much of the enormous volume of pulverized, ejected rock remained in orbit around the Earth, choking the atmosphere and blocking the sun. It may be that the asteroid's impact shook volcanoes around the planet into action, adding eruptions to the toxic mix. The smoke from worldwide fires darkened the skies across the globe. Eventually the overheated atmosphere cooled, and the blockage of the sun sent temperatures plummeting still more, creating colder global conditions than had been experienced for 185 million years—since the Permian Ice Age. Now it becomes important to be familiar with the epochs of the Tertiary period, because the first of these epochs, the Paleocene, witnessed major climactic reversals, and the next one, the Eocene, saw the beginnings of a new ice age that probably would have come whether the K/T impact occurred or not.

BACK TO THE FUTURE

As the post-impact planet cooled, shrouded in dust and smoke, there was little to suggest that an era of recovery and renewed biodiversity lay ahead. The forest fires, and the explosion into the atmosphere of huge volumes of carbonates from the impact site, greatly raised the amount of carbon dioxide in the air, creating a powerful greenhouse effect when the skies began to clear. As Figure 4-4 shows, this global warming continued through much of the Paleocene, raising temperatures even higher than they had been during the warm Cretaceous. Biogeographers conclude that this killed many plant and animal species that might have survived the blast and its immediate aftermath. But the Paleocene's warming did not continue. The next epoch, the Eocene, was marked by an almost continuous drop in global temperatures, and by the time it ended, 36 million years ago, permanent ice was beginning to form on the Antarctic continent. The Cenozoic Ice Age was about to start.

Soon the evidence began to accumulate: the early phase of the Oligocene witnessed the beginning of the formation of the Antarctic Ice Sheet even as South America and Antarctica were separating. (Remember: through all of this activity, the continents continued to move on their crustal plates, the Atlantic Oceans, North and South, kept widening, and the distribution of land and water on the planet kept changing.) Even

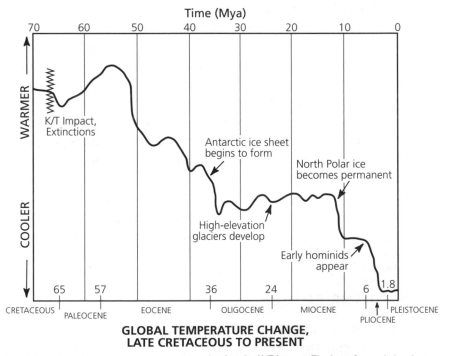

**GLOBAL TEMPERATURE CHANGE,
LATE CRETACEOUS TO PRESENT**

Fig. 4-4.   Inferred planetary temperature trends after the K/T impact. The Late Cenozoic Ice Age, still in progress, began about 36 million years ago. Hominins had to survive ever-colder temperatures and sharp climatic fluctuations during the Pliocene.

before the ice on Antarctica reached its shores, glaciers began to develop on the Earth's highest mountains, filling high-elevation valleys and sculpting a new, angular topography of sharp-edged peaks and ridges. Tree lines dropped to lower altitudes, vegetation shifted equatorward, and the mammals that had become the dominant species, including the now-common primate forms, responded by migrating and adapting as environments fluctuated.

Ice ages, as noted earlier, are not uniform cooling events: surges of coldness and advances of glaciers are interrupted by temporary warming spells long enough to reverse much of the glacial impact. So it was from about the middle of the Oligocene to the middle of the Miocene when, it seemed, the Cenozoic Ice Age had reached an equilibrium that (had weather and climate analysts been around in those days) might have been taken as a sign that the worst was over. Antarctica still had coastal zones clear of ice; the high-mountain glaciers advanced and receded as global climate cooled and warmed, and while the planet overall was cooler and drier than it had been during the ages of the dinosaurs, there was plenty of environmental variety and related biodiversity. But then, about 14 million years ago, global cooling resumed with a vengeance. Antarctica's ice sheet not only reached the ocean all around its shores, but the ice floated from land onto water and cooled the Southern Ocean, affecting the entire global ocean in the process. Permanent ice appeared and rapidly thickened on the waters of the North Pole and environs. The temperature plunge continued until a comparatively brief period of stabilization marked the end of the Miocene, but the next, brief (4.2-million-year) Pliocene epoch saw another period of further cooling. Permanent glaciers appeared even on mountains in equatorial zones of the Andes, East Africa, and New Guinea. About 1.8 million years ago began what was on average the coldest epoch of the Cenozoic Ice Age, the Pleistocene. We are living under Pleistocene conditions today, enjoying the autumn of a warm phase that, as noted earlier, has been going on for about 12,000 years.

CLIMATES AND PRIMATES

An ice age is a long-term event, lasting tens of millions of years and bringing profound changes to all parts of the planet, not just those directly affected by advancing ice sheet and valley-filling glaciers. The overall process and its manifestations may operate slowly, but there are times when sudden surges of advancing ice move fast enough to encircle grazing animals and snap off mature trees like matchsticks.

Advances of ice into lower latitudes and altitudes during a cold ice-age period are called glaciations and temporary warm-ups between glacial advances are referred to as interglacials. It should not be necessary to emphasize these distinctions, but geologists and even some physical geographers occasionally get this wrong. One of the leading geology textbooks, for example, states that "periods during which the average temperature at the Earth's surface dropped by several degrees and stayed low long enough for existing ice sheets to grow larger . . . are called glaciations (or ice ages . . .)." What should have been said, of course, is that *during* an ice age, such temperature declines are marked by glaciations (Murck & Skinner, 1999). This is important, because even while global temperatures decline during an ice age and glaciations push ice into ever-larger areas, there are also periods of relief in the form of interglacial warmth and retreating ice (a better word is *receding*, because glaciers do not really reverse direction).

Let us keep in mind, therefore, that the overall drop in temperatures that began in earnest during the Eocene and reached unprecedented lows during the most recent, Pleistocene glaciations, was no steady or continuous process. As Figure 4-4 suggests, there were long periods when the Cenozoic Ice Age seemed to have reached its maximum, with its end in sight—only to plunge into even more frigid conditions. What is certain is that the impact of that ice age was evident all over the world. When a vigorous glaciation pushed ice sheets deep into present-day Canada and into lower latitudes of Eurasia, it got cooler and drier even in Africa and equatorial South America. Tropical forests shrank, savannas expanded, and animals as well as plants responded by migrating and adapting to new environments.

We all know that the story of the great apes, hominins, and humans played itself out in Africa. All of us are, ancestrally, Africans. But any geographer looking at the world map would wonder: why Africa, when Eurasia is so much larger and seems to contain so much more environmental diversity? In that connection, one question troubled me for decades after I first learned of it in graduate school a half century ago: why are chimpanzees and orangutans, two of humans' closest genetic relatives, separated by thousands of miles of land and ocean, the chimps in Africa and the orangs in Southeast Asia?

I remember sitting in an airport van full of archeologists and anthropologists going home from a meeting in Asheville, NC about ten years ago, and raising this issue. Where, I asked, is the fossil record that would prove the migration of orangutans from Africa to Southeast Asia? It will be found, I was told. Fossils in humid South Asia don't survive like they do in Africa. We're talking 6, maybe 7 million years ago. The genetic relationship is beyond doubt, so the migration must have taken place. It's just a matter of time before the evidence appears.

Well, the evidence never surfaced, and I should have thought more carefully about the geographic implications of the question I raised. If the great apes of Southeast Asia did not descend from those of Africa, then each branch must have had ancestors in Eurasia. Postulate that, and you conclude that something drove one of those branches—the one leading to the gorilla and chimpanzee among others—to tropical Africa, while the other, leading to the orangutan, was pushed into tropical Southeast Asia.

What could have been the impetus for this dual migration? The worsening of Cenozoic Ice Age conditions, of course. Before Miocene climatic conditions deteriorated, the global environment, though cooler than it had been during the Oligocene, remained fairly stable (Fig. 4-4). During that time, the descendants of *Proconsul* and other early apes probably moved out of Africa into environmentally diverse Eurasia, where forest habitats varied widely and relatively warm temperatures ensured an ample supply of fruits and other forage. Now Eurasia, not Africa, was the heartland of differentiation and adaptation for the great apes, and numerous lineages evolved, some of them now part of the known fossil record.

But in the late Miocene the Cenozoic Ice Age took a turn for the worse, dropping global temperatures, freezing over the Arctic Ocean, drying up vast stretches of once-forested Eurasia, and destroying habitats that had for millions of years nurtured the great-ape families. Extinction was commonplace, but researchers have determined that two lineages managed to survive by adaptation and migration: *Dryopithecus*, which occupied southwestern Europe, and *Sivapithecus*, which was based in the forests of the northern Ganges River basin (Fig. 4-5).

Dryopithecus moved southward across what is today the Mediterranean into tropical and eventually equatorial Africa, probably around 9 million years ago, adapted to cope with the severe swings of environment

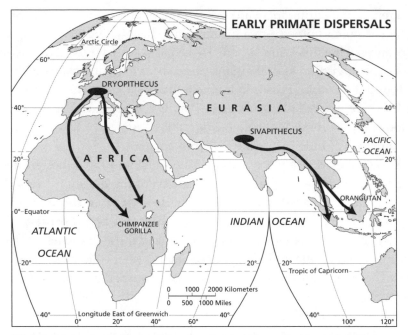

Fig. 4-5. Early primate dispersals from Eurasia into Africa and Indonesia. Late Miocene cold drove these primates southward, giving rise to the lineages of the gorilla and the chimpanzee in Africa and the orangutan in Southeast Asia. From a drawing in D. R. Begun (2003), "Planet of the Apes," *Scientific American*, 289:2.

prevailing at the time and destined to give rise to tool-making, large-brained descendants. It is somewhere along the *Dryopithecus* lineage that hominins and Africa's great apes had a common ancestor, but make no mistake: having escaped the rigors of the late Miocene in Eurasia, they found no African Garden of Eden. The increasing severity of the Cenozoic Ice Age, persisting into the Pliocene and the Pleistocene, affected tropical and equatorial Africa as well, causing rapid environmental swings that overwhelmed and extinguished numerous progenies, ape and hominin alike (Begun, 2003).

The other Eurasian great-ape lineage, *Sivapithecus*, moved down the Malayan Peninsula and into what is today Indonesia, where Miocene and Pliocene conditions may have been less rigorous than they were in ice-age Africa. In any case, no comparable evolutionary drama occurred here: the orangutan does not share a hominin ancestry and is the end of its line. Eventually and ironically, descendants of *Dryopithecus* and *Sivapithecus* would come face to face—but not ape to ape. When interglacials in the late Pliocene warmed the Earth enough to revive forests and refill desiccated lakes, Africa's hominins did what early Miocene apes had done before them: migrate out of Africa into Eurasia. And so the first of these emigrants, *Homo erectus*, moved across Arabia and southern Asia and, one day somewhere in Malaya or on Java or Borneo, saw a creature that, if the trip had been swifter, would have reminded her of the chimpanzees and gorillas left behind in Africa. Orangutan and hominin had closed a 9-million-year circle.

THE FRIGID PLEISTOCENE

Earlier we noted that the temperature plunge that began in the late Miocene and continued during the Pliocene (when early hominins made their appearance in Africa) set the stage for the Pleistocene epoch, beginning less than 2 million years ago with a series of the most severe glaciations of the entire Cenozoic Ice Age interrupted by short, warm interglacials. In Africa, where hominin evolution was under way, these phases were marked by fluctuating climates and sometimes, wild fluctuations in ecologies that guided natural selection. *Homo erectus*, successor to *Australopithecus*, was the most successful of these hominins. *Homo erectus* managed to cope with forests that changed into savannas and back again, lakes that formed and evaporated, wildlife that ranged from easy to difficult and elusive prey and, as the fossil record shows, even with massive volcanic eruptions and huge ashfalls. Undoubtedly the numbers and dispersal of *H. erectus* also varied, but the species survived for perhaps as long as 2 million years and, as we noted, left Africa and spread across Eurasia, reaching not only Southeast Asia but also present-day China in the east and Europe in the west. *Homo erectus* and its predecessor, the tool-making *Homo habilis* ("handy man"), presaged the human expansion that was to follow similar paths.

Pleistocene environmental conditions are the subject of intense investigation today. When Pleistocene glaciations were severe, permanent ice advanced deep into the landmasses of the Northern Hemisphere (Fig. 4-6). This drastically changed the distribution of plants and animals, shifting them equatorward latitudinally and downslope altitudinally. Faunal ranges and refuges shrank, niches became unusable, and always there were species that failed to survive the transition. Such glaciations could last as long as a hundred thousand years, but eventually an interglacial would warm the planet, melt much of the ice, and living space as well as survival opportunity would expand again.

Physical evidence from various sources, including Greenland ice cores and Atlantic ocean-bottom mud deposits, coupled with analyses of broken and pulverized rocks left behind by the Pleistocene glaciers, at first seemed to suggest a remarkable regularity in the ups and downs of Pleistocene temperatures. Interglacials seemed to last approximately 10,000 years, so that, on average, a glaciation-interglacial sequence encompassed around 110,000 years or so. Over the past 425,000 years of the Pleistocene, there appeared to have been four major glaciations followed by four interglacials, the latter including the current one, the Holocene.

More recent analyses suggest that it has not been so simple. First, the periodicity in the glacial-interglacial cycle seems to have changed about 425,000 years ago. Until then, each cycle took 40,000 to 50,000 years; afterward, it averaged about 100,000 years. Second, the most recent glaciation, the Wisconsinan, began about 100,000 years ago after a rather long interglacial, the Eemian. But the Wisconsinan was not one long cold spell. In fact, it was punctuated by several brief interglacials and longer (comparatively) mild spells that made habitation in higher latitudes possible for thousands of years. What is clear is that these alternations

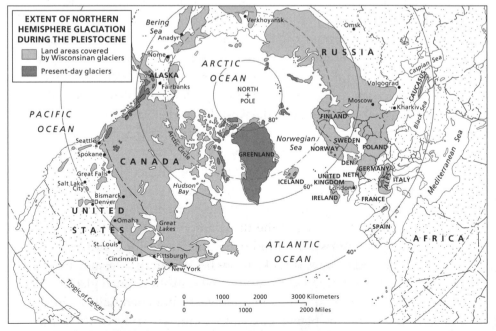

Fig. 4-6. The Northern Hemisphere as it was during a Pleistocene glaciation. Ice covered North America as far south as the Ohio River and Eurasia from Britain to Russia.

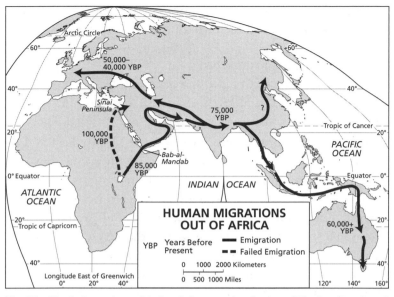

Fig. 4-7. The timing and route(s) of early human migration from Africa into Eurasia and beyond are still being reconstructed: this is one possible version.

from warm to frigid and from mild to cool often happened quite suddenly, taking their toll not only on animals and plants but also on hominins and humans.

We now begin to put modern humans in the picture, because they appeared in Africa some time during the glaciation preceding the Wisconsinan, probably around 170,000 years ago. Hominins and humans shared not only parts of Africa but also the migration routes from Africa into Eurasia, and they met similar fates when the climate took its disastrous turn. We know something about this because early humans used the land bridge between Africa and Eurasia, across the Sinai Peninsula, late during the Eemian interglacial (Fig. 4-7). No sooner were they in the Middle East when the Wisconsinan Glaciation began with a ferocious drop in temperatures. All that is left of that early emigration are the bones of the migrants. They never made it to Europe.

The next time humans tried to leave Africa they used a different exit at the opposite end of the Red Sea, probably around 85,000 years ago. The Wisconsinan Glaciation had converted so much water into ice that the surface of the Red Sea was hundreds of feet lower than it is today. The reefs in the Bab-al-Mandab (Arabic for "Gate of Grief"), where the Red Sea opens into the Indian Ocean, created the stepping stones that facilitated the crossing, and our African ancestors were on their way. First they moved along the shore of the Arabian Peninsula, then around the Persian Gulf and on into India and Southeast Asia, reaching Australia via New Guinea (another crossing facilitated by low sealevels) about 60,000 years ago (Oppenheimer, 2003).

In the process, modern humans met the hominins who had preceded them into Eurasia, and the hominins were no match for the resourceful newcomers. When the first modern humans (the Cro-Magnons, as they are known) reached Europe from India via the Middle East and found the Neanderthals, and earlier *Homo* species, on the scene, they quickly overwhelmed them with their complex culture ranging from cave art to tool kits and from inventive fishing gear to sewn clothing. They lived in cooperative communities and used sophisticated language, and thus modern humans had all the advantages over the survivors of earlier out-of-Africa migrations. Their technology gave them the opportunity to cope with the Wisconsinan's climatic swings: in milder times they expanded their frontiers, and when it got colder they devised ways to cope with increasingly harsh environments. That is why paleoanthropologists are finding fossil evidence of human settlements in Europe that survived through pretty severe cold periods. Humans were finding ways to combat the rigors of changeable climate.

A CLOSE CALL

In truth, humanity was fortunate to be able to expand into Europe at all, because the exodus from Africa via the Arabian Peninsula and South Asia almost came to a halt in spectacular fashion. We are not sure just where the vanguard of human migration was about 73,500 years ago, but the exodus across the Bab-al-Mandab was probably continuing. Then a catastrophic event in what is today Indonesia is thought to have very nearly wiped all of humanity off the planet. On the island of Sumatera (Sumatra in the old spelling), a volcanic mountain now named Toba did not just erupt—it exploded. This explosion sent millions of tons of debris into orbit, obscuring the Sun, plunging much of the Earth into long-term darkness and altering global climate. Mount Toba's detonation could not have come at a worse time. The Wisconsinan Glaciation was in full force, the Earth's habitable zone was already constricted, and a large part of the still-sparse human population faced death. Anthropologists refer to this event as humanity's "evolutionary bottleneck," suggesting that a great deal of genetic diversity must have been lost in that instant. Today, the filled-in-caldera marking Toba's cataclysm is 90 kilometers (55 miles) long and 50 kilometers (30 miles) wide, silent witness to the greatest threat to human survival since we emerged on the African savanna.

As has happened so many times in Earth's history, the skies eventually cleared, the atmosphere was cleansed, and normal conditions resumed. Toba was calamitous, but it was no Chicxulub. It did not generate fires around the world, and was not nearly as destructive as that incoming asteroid. Nevertheless, it posed a real danger to humanity, and reminds us that risks of this kind have not disappeared. Geologists refer to the Toba explosion as a 500,000-year event, something that happens, on average, once in a half-million years. That is no guarantee that another Toba-like eruption will not happen for a very long time. Our planet still poses unpredictable, incalculable natural hazards, as we were reminded on December 26, 2004.

So let us anticipate the geographic story of climate and civilization by revisiting the environmental saga of the Cenozoic Ice Age. For tens of millions of years, glaciers have been spreading during cold periods appropriately called glaciations, only to recede during warm spells we know as interglaciations. The entire planet and all its life forms, not just the polar regions, are affected by these environmental swings. Glacial advances pushed primates from cooling Eurasia into Africa and Southeast Asia; interglacial warm spells allowed hominins to leave Africa and survive in temperate Eurasian latitudes. All the while, average global temperatures declined until, about 1.8 million years ago, the start of the Pleistocene brought an alternation of long glaciations and short interglaciations. About 120,000 years ago, the Eemian interglacial was even warmer than the current one, the Holocene, but it ended abruptly some 110,000 years ago with a rapid advance of the ice of the Wisconsinan Glaciation. Humanity was on the scene by then, but our earliest emigrations from Africa were stymied by the sudden return to frigid conditions. When we finally made it out of Africa, we kept a southern route along Asian shores, but there were times when the Wisconsinan cold ameliorated and our modern ancestors took these opportunities to penetrate present-day Europe and confront—and overpower—Neanderthal predecessors. But the relentless advances of the ice came again, and just 20,000 years ago, glaciers stood as far south as the Ohio River in North America and southern England, central Germany, Slovakia, and Ukraine in Europe. Then, about 18,000 years ago, global warming sent those glaciers into fast recession, so fast that whole regions rapidly emerged from under the ice, huge ice sheets slid into the oceans, the sealevel rose, the margins of the continents were submerged, land bridges between continents and islands were inundated, and the map of the physical world began to look similar to the one we know today. Twelve thousand years ago, cold conditions made another brief comeback, but that did not last. From about 10,000 years ago until today, humanity has thrived in the warmth of a prolonged interglacial we call the Holocene, but unlike the Eemian, the Holocene has witnessed the emergence of complex cultures and civilizations, the population explosion, the formation of states and empires, the growth of megacities, and the burgeoning of industry and technology in countless forms. It has also seen wars and destruction on an unprecedented scale. With our human numbers exceeding 7 billion and "global warming" opening the last niches for habitation, the question is: what happens should the ice return, as it has more than two dozen times during the Pleistocene?

5

— — — —

CLIMATE, PLACE AND FATE

IF ONLY THE HUMANS WHO WITNESSED THE FINAL SURGE OF THE Wisconsinan glaciers could have left a written record of their struggles as the ice pushed toward lower latitudes throughout the Northern Hemisphere! There still is no conclusive evidence that humans had reached the Americas at the time, but in Eurasia, from Europe to China, it was a time of terrible environmental upheaval. As Figure 4-6 shows, sheet glaciers covered much of Europe, including nearly all of Scandinavia, Britain and Ireland, the Netherlands and Germany, Poland, the Baltic area as well as most of present-day Russia. Grinding up the rocks below and pulverizing them into fine particles that blackened their base and streaked their sides, the glaciers' relentless advance pushed piles of rubble ahead of them. Lines of trees in today's Midwest landscape mark the morainal limits of the advance of those glacial "lobes," topographic remnants of a frigid past. In North America, glacial ice reached the banks of the Ohio River. In Europe, it crossed the Rhine. In the Rocky Mountains as well as the European Alps, mountain glaciers coalesced to bury whole ranges, crushing their way down valleys and spreading onto lower plains. Plants as well as people and animals got out of the way. The coniferous forests of the north shifted southward into Spain and Mexico. The Mediterranean and the Caribbean felt the bitter cold of the Arctic. Equatorial forests in Africa, South America and South Asia shrank as savannas formed on their margins.

Twenty-thousand years ago (an eyeblink in our comparative human lifetime) there was little sign that dramatic climate change was in the offing. Even 18,000 years ago, when increasing warmth began to melt some of the glaciers on their margins, no scientists would have forecast that the ice would continue to recede, or that the glaciers would disappear in short order. And the people experiencing it probably thought that this would be just another temporary respite of the kind human communities surviving in high latitudes had experienced before. But this time the global warming that had started turned out to be something different. It was so powerful and persistent that the glaciers began to give way, dropping their pulverized load and exposing the ground below—not just at their leading edges, but all over the landscape.

It must have been a turbulent time. Huge volumes of meltwater streamed from the fronts of the glaciers as well as from their interior surfaces, generating roaring rivers choked with sediment much of which filled the Mississippi Valley and Delta. While sea level was still low, raging rivers like the Hudson carved deep canyons in the continental shelf. But sea level would soon rise and drown these products of the Great Melt, evidence of physiographic turmoil around the world. In Europe, the rising water severed Britain from the mainland as the English Channel formed and widened. Meltwater poured from the glaciers on the Scandinavian Peninsula and filled the Baltic Sea. Ice fast disappeared from Germany and Russia. The glaciers of the Alps receded as the mountains' ice cap thinned under the hot sun.

To our Stone Age ancestors who had been living with the late Wisconsinan's frigid climate, this warming must have been a welcome but also a challenging experience. Accompanying the warming was environmental chaos—as the glaciers released their pulverized, powdered rock, powerful, storm-strength winds blew the stuff (called *loess*) into the air in choking clouds that made the Dust Bowl in the American Great Plains of the 1930s child's play by comparison. In both North America and Eurasia this loess settled in thick layers many meters deep, and it is not surprising that a map of the most recent deposits shows considerable association with the rivers fed by melting glacial ice such as the Huang (Yellow) in China, the Danube in Europe and the Mississippi-Ohio drainage in North America. If anthropologists are correct in their hypothesis that humans did not reach the Americas much more than 13,000 years ago, no human communities here had to deal with this periglacial ("edge of the glaciers") phenomenon. But there is no doubt that humans in Europe and Asia had to deal with it. Nevertheless, the warming was so persistent that people ventured farther and farther poleward.

Warmer summers and milder winters opened opportunities for more secure livelihoods and larger communities.

ONE FINAL SURGE

But nature still had a surprise in store. By about 12,000 years ago, the ice had melted from southern regions in both North America and Europe, but still held on in what is today Canada and parts of northern Europe and northern Russia. Because these ice sheets were lubricated, even at their bases, with meltwater, the inevitable happened: they started to slide downslope, that is, coastward and into the ocean. One especially large ice sheet, the size of a large Canadian province, slid into the North Atlantic, causing disastrous waves along coastlines from Europe to the Caribbean and chilling the ocean right back to glaciation-time temperatures. This event, called the Younger Dryas after a Tundra wildflower that flourishes where the ice once stood, caused more than a thousand years of cooling (our ancestors must have said "here we go again"), but this time that cooling was temporary and brief. By about 10,000 years ago, temperatures were back on the global-warming track, the remaining ice was melting again, and the Younger Dryas was just a hiccup in the overall pattern.

We should note that the primary hypothesis—that a huge piece of northern ice slid into the North Atlantic and caused the Younger Dryas—is being challenged today by scientists who postulate that some cosmic event, possibly a comet exploding close to the surface somewhere over the Northern Hemisphere, caused atmospheric conditions that resulted in temporary cooling. No consensus has been reached, but there is a lesson in this: a long-term temperature trend can be quickly reversed by an environmental incident. What would happen if the already-receding Greenland Ice Sheet suddenly slid *en masse* into Arctic waters? Or if a significant portion of the Antarctic Ice Sheet collapsed into the Southern Ocean? Well, we wouldn't have to worry about global warming for a while.

HOLOCENE HUMANITY

The post-Younger-Dryas warm period that has now lasted some 10,000 years is referred to by geologists as the Holocene Epoch, but of course there is no evidence as yet that this is geologically a new epoch, that is, a

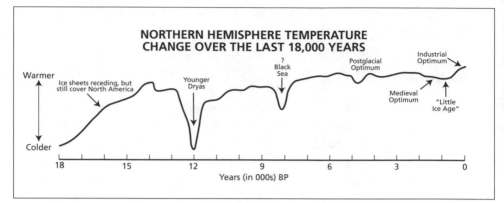

Fig. 5-1. Temperatures in the Northern Hemisphere over the past 18,000 years. Rapid warming and fast melting of the Wisconsinan glaciers is followed by a sudden return to bitterly cold temperatures (the Younger Dryas) but just as quickly the warming trend resumes and has continued, with some fluctuations, until the present day. Today, after a cooling period from about 1940 to 1970, global warming is augmented by human pollution of the atmosphere.

new phase following the end of the Cenozoic Ice Age. All the Holocene is, so far as can be known, another Pleistocene interglacial.

What *is* different about the Holocene is not geologic, but geographic. About the fact that this is a completely new *cultural* epoch there can be no doubt: hominins have been around for millions of years, modern humans maybe 170,000, but nothing like the demographic or cultural explosion witnessed by the Holocene has ever happened before. In 2011 some scholars proposed that the geographic equivalent to the geologic Holocene should be called the Anthropocene, in recognition of the impact humans have had on the planet and its environments. In support of this notion, David Deutsch points out that (for example) the present-day urban topography of the Island of Manhattan completely obliterates and overpowers the geologic one, making the geologic landscape essentially irrelevant and the anthropogenic topography totally and functionally dominant (Deutsch, 2011). Those who hypothesize that human pollution of the atmosphere is the sole cause of the recent average increase in planetary temperature point to humanity as the dominant environmental agent, further justification for the "Anthropocene."

But the notion of human dominance over nature during the past ten thousand years remains little more than that. True, the past ten thousand years have witnessed the conversion of vast reaches of natural vegetation into farmlands, the damming of rivers, the irrigation of hillslopes, the submergence of urban sites beneath asphalt and concrete, the extraction

of natural resources, the near-destruction of marine fauna, the deforestation of much of the tropics, and the extinction of countless species—all
this while the human population grew from a few tens of thousands to
7 billion. But nature's power remains awesome. A solar blast of a few
billion tons of protons through the Earth's magnetic cocoon would fry
the whole cyberelectrosphere of modern electronics. A volcanic eruption
like that of Toba little more than 70,000 years ago would cost millions of
lives and set the Earth back for decades. Just a couple of earthquakes on
December 26, 2004 off Sumatra generated a tsunami that cost more than
a quarter of a million lives. An asteroid impact comparable to Chicxulub
would put paid to—well, the Holocene/Anthropocene.

Certainly the postglacial warm-up that led to today's mild planetary
temperatures still holds its secrets. One of the continuing points of scientific research (and debate) has to do with the Black Sea, which apparently filled up rapidly about 8,200 years ago. Some years ago, a team of
geologists hypothesized that this, too, was the result of a last mass of ice
sliding off northernmost North America, smaller than the one assumed
to have caused the Younger Dryas but big enough to lower North Atlantic temperatures and raise sealevels. Those raised Atlantic sealevels
caused water to pour through the Strait of Gibraltar into the Mediterranean Sea, and in turn the Mediterranean overflowed its barrier with
the lake that was then the Black "Sea," on whose shores stood numerous
villages. Geologists William Ryan and Walter Pitman described this event
as having the force of 200 Niagara Falls, filling the Black Sea at a rate of
15 centimeters (6 inches) per day and forcing the coastal villagers to back
away about 1.6 kilometers (1 mile) daily. Many of them abandoned their
dwellings, boats, fields, and even animals and watched the flood swal-
low them up. By the time it was over, Ryan and Pitman wrote, the surface of the Black Sea had risen 145 meters (500 feet)—and, quite possibly,
the biblical legend of the Great Flood was born.

Subsequent research has cast doubt on this scenario, but there is no
doubt that something dramatic happened in the Black Sea basin for
which the locals were not prepared. Other dramatic events elsewhere
in the world are probably awaiting discovery and analysis. In any case,
the Holocene was no uneventful transition from glacial to warmer and
calmer times. Still, conditions did stabilize enough following the postulated Black Sea crisis to allow climatologists to talk of a Postglacial
Optimum, starting about 7,000 years ago and marking the time when
global environmental conditions became rather like those with which

we are familiar today (Fig. 5-1). This Postglacial Optimum transmuted into another salubrious period, the Medieval Optimum, starting around 2,000 years ago, which witnessed the expansion of settlement in northern latitudes of Eurasia, the peopling of Iceland, even the colonization of Greenland.

Not all was quiet during these climatically optimal times. About 3620 BP (Before the Present) the volcanic island of Thira (Santorini), to the north of Crete in the Greek archipelago now known as the Cyclades, blew up in a Toba-like eruption that covered a wide area of the eastern Mediterranean with a thick layer of poisonous ash. Days of darkened skies and seismic waves in the waters may have given rise to biblical allusions to darkness and parting waves (and perhaps to the legend of Atlantis, for when daylight returned, Thira was mostly gone). But more consequential is what happened on Crete, where the powerful and culturally advanced Minoan civilization was based. Thira may have dealt it a fatal blow, opening the way to the regional dominance of ancient Greece.

Nor was this optimal period a time of invariably favorable climate. De-glaciation (the recession of ice and the opening of land to atmospheric conditions) continues long after the ice has disappeared: the poleward shift of climatic zones, the maturing of soils, the migration of plants and animals keep altering the environment for thousands of years. Some societies found themselves in favorable locales and converted their good fortune into security, expansion, and power. Others, including some early states and cities in what is today the Middle East, saw rivers dry up, deserts encroach, and livelihoods destroyed. Was the innovation of irrigated farming a response to these challenges, as some scholars suggest, or had population pressure in stable environments led to it previously?

The Medieval Optimum certainly was a good time for two contemporaneous empires: the Roman in western Eurasia and the Han in eastern Eurasia. The Roman Empire unified Europe as never before (or since) and put an indelible cultural stamp on much of it. The empire of Han was China's formative dynasty and laid the foundation for a large and powerful state. The Han capital, then called Ch'angan (now Xian), was known as the Rome of China; Rome was the Ch'angan of the Mediterranean. The Silk Route not only transferred goods between East and West, but also tales of splendor and power.

And it was warm, in western Europe as well as in eastern Asia. The Romans planted grapevines in Britain and left behind a thriving wine industry. The agricultural frontier moved steadily northward in Scandinavia, and treelines and pastures moved upward on Alpine slopes. Meanwhile Europe's medieval cities mushroomed as architects endowed them with some of the civilization's major works: the Notre Dame Cathedral on an island in the Seine River in the heart of Paris; the cathedral at Chartres, the cathedral at Canterbury, and numerous other triumphs of Gothic styling and engineering. In China, the Tang Dynasty (618–907) brought a golden age of expansion and consolidation, architecture and art. Superbly designed pagodas heralded the diffusion of Buddhism; Xian was the cultural capital and the largest city in the world. The ensuing Song Dynasty benefited from the unprecedented agricultural productivity of the North China Plain and the rice fields to the south: by the end of the dynasty, 1279, China had an estimated 100 million inhabitants. And when the Mongols invaded China to establish their Yuan Dynasty they, too, benefited for some time from the beneficent conditions of the Medieval Optimum.

But not for long. Had the Chinese and the Europeans been exchanging weather information the way we do today, the Chinese would have been alarmed at what they heard from Europe. Because in the west, the mild and pleasant conditions prevailing for so long that they had become the accepted norm showed signs of ending. Winters got colder, May frosts, hardly known for centuries, became common. Early fall frosts led to local famines. Persistent droughts hit some parts of Europe; destructive floods struck elsewhere. Britain's wine industry was erased by cold in a matter of decades. By the turn of the fourteenth century, Alpine glaciers began to advance. Greenland's small settlement had long since disappeared; Iceland was abandoned as well. Weather extremes abounded, not only in the form of record cold snaps but also as searing summer heat and raging storms. Nature seemed to be preparing for one of those climatic reversals so common during the Pleistocene. Was the Holocene itself coming to an end?

THE LITTLE ICE AGE

To the farmers, winegrowers, and seafarers of the fourteenth century it must have seemed so. Increasing cold, decreasing rainfall, frigid winds, and shortened growing seasons made for dwindling harvests, failing farms, and seas too stormy for fishing. Famines struck all over Europe, just at a time when more people were clustered in towns than ever before. And the climatic record, pieced together from farmers' diaries (those of winegrowers are especially useful), tree ring research (dendrochronology), ice cores, contemporary writings, illustrative paintings, and surviving sketches and drawings, justify the designation of the post-1300 period as a reglaciation. We now know that this return to colder times, marked by advancing mountain glaciers and thickening Subarctic ice, would end in the mid-nineteenth century, and that even the worst of it, starting in the late 1600s, did not lead to full-scale Pleistocene glaciation. Whatever was happening caused havoc in Europe, and in other parts of the world as well, but of course those who experienced it were unaware of the long-term implications. Only when new methods of analysis became available did scientists realize what had happened—and then they gave the episode an inappropriate name. This temporary cooling was no ice age: it was a minor glaciation, and not the first over the past 7,000 years. But the name "Little Ice Age" certainly was more dramatic than "Minor Glaciation," and it stuck.

To those affected, it was anything but little or minor. Europe's climate fluctuated wildly, often suddenly, so that recovery would be followed by renewed famine, as populations mushroomed and then collapsed again. In the fourteenth century, the geography of eastern and western Eurasia became fatally interlocked. The salubrious conditions that enabled Mongol peoples to thrive and expand, leading to a Mongol dynasty in China, also energized their penetration westward. In the process, Mongol migrants and their horse caravans picked up the strain of bacteria that brings on the bubonic plague, and its vector, the flea, rode into Europe on rats and people ahead of their relentless advance. The Black Death swept over an already weakened Europe in waves that often killed half the population or more, and recovery, medical as well as environmental, did not start until the last quarter of the fifteenth century.

In China, meanwhile, the full impact of the Little Ice Age occurred after the end of the Mongol (Yuan) Dynasty (1368). The early Ming rulers inherited a populous state sustained by wheat in the north and rice in the center, linked by the Grand Canal and other busy waterways. Late in the fourteenth century the Ming rulers, exhorted by the legendary admiral Chung Ho, authorized the construction of an ocean-going fleet that would stake China's claim, and enhance its reputation, in the Indian Ocean and beyond. The fleet eventually numbered more than 6,000 ships, the largest carrying as many as 500 men; these were 400 feet long, had four decks and nine masts, and carried sufficient fresh water and supplies to sail for 30 days. Nothing built in Europe even began to approach these vessels in terms of technology or capacity, and the first expedition, in 1405, involved 315 ships and 27,000 men. Later voyages reached the Persian Gulf and the Red Sea as well as East Africa, possibly as far south as Sofala. The Chinese seemed poised to round the Cape of Good Hope and enter the Atlantic.

But then disaster struck at home. The first onslaught of the Little Ice Age came later than it did in Europe, but it was no less severe. Interior rains failed, rivers dried up, the wheat crop shrank, famines broke out, and social disorder and epidemics raged. The Ming rulers ordered an end to the maritime expeditions, dictated the burning of all ocean-going vessels, and instructed the Nanjing shipyard, then by far the largest in the world, to build only barges that could navigate the Grand Canal with cargoes of rice, thus to alleviate the plight of the colder, drier north. Environment may not determine the capacities of humans, but environmental events can decisively influence the course of history.

CRISIS IN EUROPE

It is no exaggeration to say that the vicissitudes of Western Europe's Little Ice Age climate brought about what cultural geographers refer to as the second Agricultural Revolution. Farm implements were improved; field methods (planting, sowing, watering, weeding, harvesting) got better, transportation and storage of produce involved less waste and loss. New crops were tried (not always with good results); marketing in the growing urban areas became more efficient. All this was, literally, a matter of survival, because toward the end of the sixteenth century there were signs that the Little Ice Age had even worse in store. The century closed with one of the most extreme decades in Europe's known environmental history, and during the seventeenth, conditions were worsened by a series of volcanic eruptions in Southeast Asia, precipitating colder spells in an already frigid region (Grove, 1988).

Some climatologists call the period from about 1650 to 1850 the "real" Little Ice Age, citing the environmental crisis that gripped Europe during those 200 years as much more severe than anything that had gone before. And indeed, between 1675 and 1735 the planet appears to have experienced the coldest cycle of the millennium. Growing seasons in parts of Europe were shortened by as much as six weeks; ports were blocked by ice; the Denmark Strait between Iceland and Greenland remained ice choked and impassable even during the summers. Sea ice formed and remained in place over the North Sea as far as 35 miles (55 km) from shore. Priests and their parishioners prayed at the faces of fast-advancing Alpine glaciers threatening villages and farms.

Referring back to the complicated interactive cycles that drive climate change, it is worth noting that this coldest period of the millennium (1675 to 1735) appears to coincide with a probable phase of minimal Sunspot activity called the Maunder Minimum after the scientist who analyzed it. The qualifiers are necessary because Sunspot activity records before about 1750 are incomplete, but the correlation has become a topic of considerable interest because little is known about solar cycles long term. Another, better defined solar minimum, the Dalton Minimum, marks the period between about 1795 and 1820, when once again it was extraordinarily cold, and not only in Europe.

Was this indeed a global phenomenon? In his book *The Little Ice Age* archaeologist Brian Fagan describes how the Franz Josef Glacier on New Zealand's South Island "thrust downslope into the valley below, smashing into the great rainforests . . . felling giant trees like matchsticks" (Fagan, 2000). In North America, our growing understanding of the Little Ice Age helps explain why the Jamestown colony collapsed so fast, a failure attributed by historians to ineptitude, lack of preparation, and racist attitudes toward the Native Americans in the area. The chief cause probably was environmental. Geographer Don Stahle of the University of Arkansas and his team, studying tree ring records that go back eight centuries, found that the Jamestown area experienced a seven-year drought between 1606 (the year before the colony's founding) through 1612, the worst in nearly eight centuries. European colonists and Native Americans were in the same situation, and their relations worsened as they were forced to compete for dwindling food and falling water tables. The high rate of starvation was not unique to the colonists. They, and their Native American neighbors, faced the rigors of the Little Ice Age together.

In Europe, there was little respite. The decade of the 1780s brought one crisis after another. A gigantic volcanic eruption on Iceland in 1783 lasted for eight months and ejected an estimated 100 million tons of ash, sulfur dioxide, and other pollutants into the atmosphere. The Laki eruption lowered temperatures in North America by 7 degrees Fahrenheit, and brought on a series of frigid winters in Europe, Russia, and even North Africa. In February 1784, ice blocked the entire lower Rhine River,

producing the worst floods in recorded history and causing food shortages and general economic distress. Violent weather in Western Europe
in 1788 included hailstorms that felled forests (one report refers to hail-
stones 15 inches in diameter) and storms that flattened crops. The French
Revolution had other causes, of course, but its timing undoubtedly was
related to the food shortages recurring during that dreadful decade. Nor
was Napoleon Bonaparte fortunate in his timing when he marched
against Russia in 1812. The period from 1805 to 1820 was one of the coldest in the "real" Little Ice Age, and when Napoleon's armies invaded
Russia their biggest adversary was the bitter winter, with which Russian forces were more familiar.

DISTANT THREAT

Just when it must have seemed that conditions could not get any tougher,
they did—not because of an atmospheric event but as a result of a volcanic eruption on the other side of the planet.

On April 5, 1815, the Tambora Volcano on the island of Sumbawa in
what was then the Dutch East Indies, located not far east of Bali, rum-
bled to life. Less than a week later it was pulverized in a series of explosions that could be heard a thousand miles away, killing all but 26 of the
island's population of 12,000. When it was over, the top 4,000 feet of the
volcano were gone, and most of what is now Indonesia was covered by
debris. Darkness enveloped much of the colony for weeks, and tens of
thousands died of famine in the months that followed. Colonial reports
describe fields covered by poisonous ash and powder, waters clogged by
trees and cinders, air rendered unbreathable by a fog of acid chemicals.

Tambora's explosions rocketed tens of millions of tons of ash into
orbit, darkening skies around the world. What began as a narrow equatorial band of ash and dust gradually widened into a globe-girdling
membrane that blocked part of the Sun's radiation. By the middle of 1816,
it was clear to farmers everywhere that this would be a year without
summer, a growing season without growth. In Europe, food shortages
were acute and grain prices rose rapidly, forcing governments to close
their borders to prevent speculation. Food riots nevertheless broke out in
the towns, and in the countryside armed gangs raided farms and stores.
In the United States, the "year without summer" was especially difficult
on the farms in New England, where corn would not ripen, grain prices
escalated, and the livestock market collapsed. We can only guess at the

impact of Tambora's eruption in other parts of the world, but there can be no doubt that 1816 was a desperate year in the Little Ice Age—a crisis that reminds us of the risks under which all of humanity lives.

THE HUMAN FACTOR

To put these chronological events in spatial perspective, we should remind ourselves of what was happening to the human world as the Little Ice Age came to its mid-nineteenth-century end. The Industrial Revolution was gathering steam; the colonial era was transforming societies and economies from Central America to Southeast Asia. Europeans were populating and dominating distant lands, fighting among themselves even as they exploited their imperial domains. And population growth was accelerating. Around the time of the Tambora eruption, the Earth's population was about 1 billion, perhaps twice what it was at the beginning of the Little Ice Age. Since then, a human population explosion has coincided with the post–Little Ice Age warming that has been in progress, with one major interruption from approximately 1940 to 1970, since about 1850. Consider this: by 2015, two centuries after Tambora, the Earth will carry seven times as many people as it did when that volcano exploded. How would the world cope today with a "year without summer," not to mention a Toba catastrophe—or a sudden return to Pleistocene glaciation?

WARMING THE WORLD

Overshadowing any prospects of long-range environmental challenges based on Pleistocene history is the current reality: the Earth is in a warming phase that, if sustained over centuries, has the potential escalate into a global crisis. When the great global warming began that cleared the planet of most of its Wisconsinan-glaciation ice (the Antarctic and Greenland ice sheets now form the last large remnants of this episode), the continental shelves adjoining many coastlines rapidly flooded as sea level rose relentlessly. There can be no doubt that human communities were driven from their homes when this happened. Had cities been built along the glacial-period shorelines, they would today be under water. Now the possibility arises that further global warming, speeded up by human pollution of the atmosphere, will cause a further rise in sea levels. Estimates of the potential amount of this rise vary, and it depends, of course, on how much of the remaining ice will melt, and how fast. A total melting of both the Greenland and Antarctic ice caps would produce an unthinkable sea-level rise of many meters, drowning New York, London, Mumbai, Shanghai and hundreds of other cities and towns. Since it does not appear that such complete melting happened during previous interglacials, this may be unlikely—but on the other hand, human action wasn't part of the earlier process.

Even a less dramatic rise in sea level will have a fateful impact on vulnerable countries. Populated clusters of low-elevation islands in the Pacific and Indian Oceans have become independent nations, and any significant rise in sea level threatens not just their margins, but their very existence. The highest land in such states as Tuvalu in the Pacific and the Maldives in the Indian Ocean lies less than 6 meters (20 feet) above sea level (7 feet in the Maldives), so options for moving to higher ground do not exist. So concerned over the future are some governments in these countries that they have threatened to sue industrial states for enhancing the greenhouse effect and have sought land grants from neighbors on which to relocate the entire population in case of inundation. Scientists' dire predictions can have unanticipated consequences.

So let us take stock of the current situation in context of what has happened during the last quarter of the Pleistocene Epoch, the last 425,000 years of climate change that, with research in many branches of science, can be reconstructed with considerable confidence (Fig. 5-2). Note that this period has witnessed five interglacials (warm periods of the kind we are experiencing today) and four much longer glaciations marked by major temperature swings and advancing and receding ice. These four glaciations are named as the Nebraskan, Kansan, Illinoian and Wisconsinan (they are matched by four European glaciations with different names). Between the Illinoian and Wisconsinan glaciations came the Eemian Interglacial, experienced but not affected by our distant human ancestors, and after the Wisconsinan glaciations with its dramatic final surge of the ice—watch the dropping temperatures on Figure 5-2—came "our" interglacial, the Holocene.

Obviously the Eemian interglacial is of much interest, because it may tell us something about what to expect in our own time. Notice the fast temperature rise that ended the Illinoian Glaciation, the warmth of the early Eemian Interglacial, and then the decline in temperatures leading to the Wisconsinan Glaciation. The climate scientist James Hansen, whose brilliant analyses of these events did much to temper the

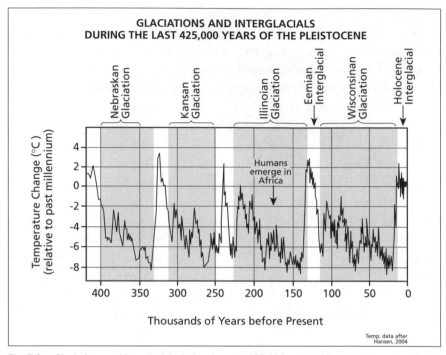

Fig. 5-2. Glaciations and interglacials during the past 400,000 years of the Pleistocene. Note that glaciations are not uniformly cold and interglacials are not consistently warm. Humans settled Europe during some of the Wisconsinan's coldest times. Modified from J. Hansen (2004), "Defusing the Global Warming Time Bomb," Scientific American, 290:1.

"global-warming" debate, has pointed out that the highest temperatures during the Eemian were even warmer than we have seen during the Holocene, but the human role in enhancing the greenhouse effect may eventually make the Holocene warmer still. Hansen's calculations are troubling: during the Eemian, sea levels rose as much as 3 to 5 meters (10 to 16 feet) above the present level (Hansen, 2004). Obviously, interglacials do not reach the same level of warmth every time: the evidence suggests that the interglacial separating the Nebraskan and Kansan Glaciations was warmer still, but the one after the Kansan Glaciation never got as warm as any of the others.

So we may be about to experience the highest temperatures of the entire Holocene, combining natural causes and human intervention to create a new maximum—or nature may have in mind what it did before, ending our interglacial with precipitous cooling. At the moment, there appears to be no prospect of the latter, and the obvious wild card is human action. To be sure, there is no doubt that humanity's prodigious emanations of greenhouse gases is contributing significantly, perhaps dominantly, to the planet's current warming. The question is whether these actions have the capacity to overpower nature's cyclic reversals, so that the steep temperature drops seen to end previous interglacials on a fairly regular pattern may not be repeated in the later Holocene. Somehow it seems arrogant to assume that we have the ability to subdue nature (given what nature has overcome after the massive pollution following Chicxulub, Toba, and other catastrophic events), but some scientists hypothesize that we may be heating the planet forever, ending the sequence shown on Figure 5-2. If they are proven correct, it will give new meaning to the term "Anthropocene."

INDUSTRIAL OPTIMUM

The environmental events of the past two centuries, that is, following the end of Europe's "Little Ice Age," may well be called the Industrial Optimum given the impact humans and their works have had on the atmosphere ever since the start of the Industrial Revolution. A chart showing average global temperatures from the middle of the nineteenth century to the second decade of the twenty-first, is a record of overall warming—except during what at the time was the height of the Industrial Revolution, approximately 1940 to 1970 (Fig. 5-3).

This mid-twentieth-century reversal gave rise to three decades of scientific analysis and prediction, and the scholarly as well as the popular literature was full of forecasts of impending glaciations. Personal recollections of this period still survive: my diaries describe the bitter cold and extraordinary length of winters during the Second World War in the Netherlands, then some frigid winter times in Johannesburg, South Africa in the late forties, followed by September snowfalls in Evanston, Illinois in the fifties and May frosts in East Lansing, Michigan in the sixties. As a graduate student at Northwestern I remember a dramatic lecture by a visiting British scholar who not only predicted the impending end of the current interglacial but who had rigged up a device that started dropping "snow" on him and his lectern as he concluded his talk. "Global cooling" was the mantra in those days, and dissent was as difficult then as it is now. Have a look at scientific journals such as *Science* and *Nature* from that period and you will see numerous articles propounding global-cooling hypotheses and almost none espousing an opposite or cautionary view.

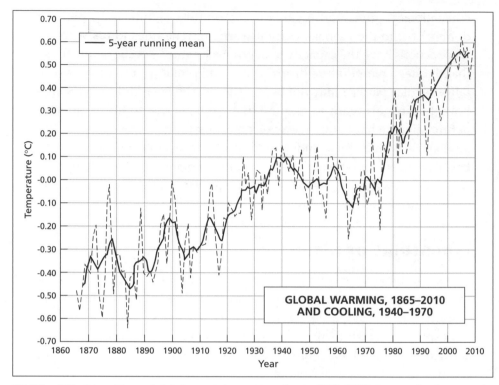

Fig. 5-3. Aggregate evidence of global average temperature change, 1860–2010. A composite of many such calculations, this record suggests a rise in temperature in excess of 1° centigrade. Note the average decline, 1940s to 1960s, and the accelerated rate of warming after 1970.

These articles became fodder in the global-warming debate decades later (the columnist George Will used them relentlessly to cast doubt on the current consensus), and scientists revisited the issue later in search of an explanation for this "inconvenient truth." Their conclusion was that a 1940s change in industrial technology had resulted in massive emissions of sulfurous gases, whose effect was to reflect the Sun's radiation back into space, thus causing cooling; the resumption of warming in the 1970s, in this hypothesis, evinced the effect of the ever-growing volume of greenhouse gases being poured into the atmosphere. While the science appeared sound, critics asked how such global cooling could have occurred just when the industrializing Soviet Union's prodigious output of greenhouse gases was adding to global emissions. Meanwhile, the cooling phase and its tough winters had practical impact on the ground: in the United States, many people in the Midwest and Northeast responded by moving southward either permanently or seasonally (the "Sunbelt" thus became a geographic appellation).

Since the mid-1970s, and especially during the past two decades, average global temperatures have continued to increase, and the consequences, ranging from shrinking Arctic sea ice and the prospective opening of northerly maritime routes to the spread of disease-carrying insect vectors into areas once immune, are omnipresent. It is sometimes said, these days, that while "global warming" is a threat to most peoples and societies on this planet, there will be winners as well as losers. The winners might include those living in high, cold latitudes where warmth will shorten winters and brighten summers. Russia, in particular, might see gains from global warming: the recession of Arctic ice would open submarine energy reserves to exploitation, would free ice-bound ports for navigation, and would soften the harsh environments of Siberia. But Russia's hopes were tarnished by what happened in 2010, when forest fires during searing summer heat destroyed whole villages, killed more than 50 people, left thousands homeless and enveloped the capital, Moscow, in a poisonous smog. Climate change, short- or long-range, rarely comes without a price tag.

ABRUPT CLIMATE CHANGE AND DANGEROUS EXTREMES

Much evidence indicates that climate change can occur rapidly and that short-term reversals are likely to be attended by violent environmental events. You do not have to be a dedicated follower of the daily news to learn of the troubling incidence of extremes recorded worldwide, from Russia's desiccating 2010 summer to New Zealand's frigid 2011 winter, from torrential rains in desert areas to parching droughts in normally moist environs, from early autumn frosts to late blizzards, from stronger hurricanes to unusually frequent tornadoes. If we could interview a European climatologist from the 1300s, we might well be told that it all seems familiar. Climatic reversals seem to be presaged by exceptional weather events.

Will we, in our enormous numbers, ever be able to adapt to rapid environmental change and save ourselves from the chaos that attended the onset of the Little Ice Age? As Brian Fagan states, it may be an illusion to suggest that humanity, through its technological prowess, will ever be able to adjust to the kinds of changes nature has on the record.

"Climate change," he writes, "is almost always abrupt, shifting rapidly within decades, even years . . . it is unpredictable, and sometimes vicious. The future promises violent change on a local and global scale . . . such cycles are frightening to contemplate in an overpopulated and heavily industrialized world" (Fagan, 2000).

Scientists are zeroing in on the question of abrupt climate change, and they are finding that the current global warming is likely to trigger rapid environmental shifts that could cause chaos on the planet. Ice cores from Greenland, nearly two miles long and revealing a climatic record spanning the past 110,000 years, record wild fluctuations including slow warming and sudden cooling, the latter (before the Little Ice Age) sending icebergs as far south as the coast of Portugal (Alley, 2002). High-latitude cold had regional effects all over the world. For example, about 5,000 years ago a northern cold spell coincided with the drying of the Sahara astride the Tropic of Cancer, converting it from a verdant landscape with rivers and lakes to the parched, rocky, and sandy wasteland it is today. That desiccation did not happen over millennia—it occurred over just a few decades, with far-reaching impact on the entire continent's human geography. The Greenland ice cores show that nature's variability is endless and restless. Now, scientists theorize, the additional impact of human activity on the global atmosphere may trigger even more sudden climate change than prevailed long ago. Sooner or later we will face extremes that come upon us quickly and will give us little time to find ways to cope with the consequences. For all our technological prowess, we still depend on nature to sustain us.

We are living, in the words of the geneticist Stephen Oppenheimer, in the autumn of the interglacial Holocene, the brief epoch that has witnessed the transformation of our human world from small villages to megacities and from simple to complex cultures, from isolated communities to interconnected empires and from stone tools to spacecraft. On the scale of our lifetimes, it has all happened in the past few seconds, and our modern experience with the kinds of challenges nature can pose—from meteor impacts to glaciations—is virtually nil. We will never be able to control climate change, but we may be able to mitigate it somewhat by limiting our greenhouse-gas emanations. And we should begin planning for worldwide coordination in the event of global natural emergencies caused by nature, of which we already have been amply warned. The impact of an episode of rapid climate change poses as great a potential challenge to this nation as any it will face in the years ahead.

CLIMATE AND WEATHER ON THE MAP

Our focus on climate change should not obscure the reality that most of us tend to be concerned about the weather on a daily basis. The difference is obvious: the term "climate" represents an overall picture of the year-round conditions prevailing at a given place. It is an average of all available data, especially temperature and precipitation and their seasonal ups and downs. That's why we speak of a tropical climate or an Arctic climate. But we don't pick up the morning paper or check the Weather Channel for the climate: it's the daily *weather* we're interested in. This, of course, refers to the conditions at hand. We want to know the temperature, humidity, wind strength and direction and what the day will bring. If it is 90 degrees Fahrenheit and raining with high humidity and little wind, that is the weather—probably somewhere in a tropical climate.

All the weather-making factors combined—the sweep of the Sun's energy, the rotation of the Earth, the circulation systems of the oceans, the movement of pressure systems, the rush of air in jet streams—produce a global pattern of climates that may look complicated at first but is actually remarkably simple. This is one of those maps that is worth a million words, because it allows us, at a glance, to determine what the prevailing climate is anywhere on Earth, and what we may expect in the way of weather under those climatic conditions (Fig. 5-4).

We owe this remarkable map to the work of Vladimir Köppen (1846–1940), who devised a scheme to classify the world's climates based on indices of temperature and precipitation. Since his time, there have been many efforts to improve on Köppen's classification and regionalization of climates, but nearly a century after its creation, and despite the greater accuracy of climatic data today, it has stood the test of time (Fig. 5-4).

Using not only climatic but also biotic information, Köppen established six major types of climate. He used letters to identify these on his map:

> *A: Equatorial, Tropical, Moist*
> *B: Desert, Dry*
> *C: Midlatitude, Mild*
> *D: Continental, Harsh*
> *E: Polar, Frigid*
> *H: Highland*

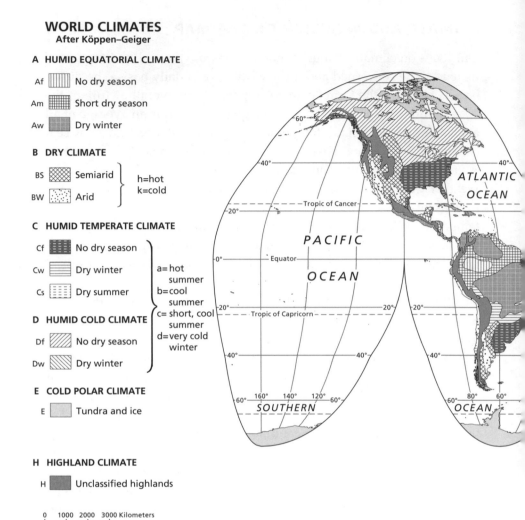

WORLD CLIMATES
After Köppen–Geiger

A HUMID EQUATORIAL CLIMATE

Af ▥ No dry season

Am ▦ Short dry season

Aw ▨ Dry winter

B DRY CLIMATE

BS ▨ Semiarid ⎫
⎬ h=hot
BW ▒ Arid ⎭ k=cold

C HUMID TEMPERATE CLIMATE

Cf ▦ No dry season ⎫
⎪
Cw ▤ Dry winter ⎪ a=hot
⎪ summer
Cs ▥ Dry summer ⎪ b=cool
⎪ summer
D HUMID COLD CLIMATE ⎬ c= short, cool
⎪ summer
Df ▨ No dry season ⎪ d=very cold
⎪ winter
Dw ▨ Dry winter ⎭

E COLD POLAR CLIMATE

E ▢ Tundra and ice

H HIGHLAND CLIMATE

H ▪ Unclassified highlands

0   1000 2000 3000 Kilometers
|——|——|——|
0   1000   2000 Miles

Fig. 5-4. The Koppen-Geiger map of world climates. Note the vast expanses of desert and steppe, rainforest and polar climates; it is interesting to compare this map to Figure 3-2, World Population Distribution. Our planet is small; most of its surface is water and ice; only a comparatively small part of the land is habitable with adequate water and moderate temperatures—and yet 7 billion of us are aboard.

Köppen added other letters to provide more detail within each of these climatic regions, but even without these, his map is very useful. Take, for example, the climate prevailing over most of the southeastern United States, Cf on the map. If you're familiar with that climate and its local weather (for example, in Atlanta, or Nashville, or Charlotte), you'll have

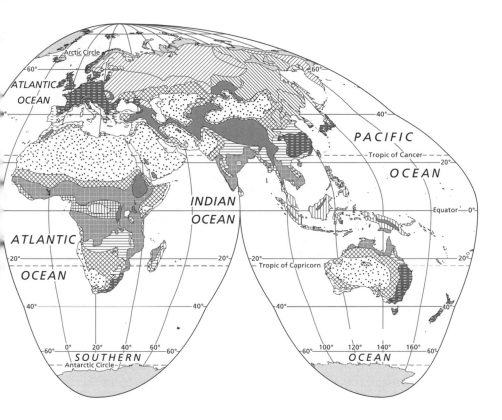

a pretty good idea of what it's like in much of eastern China, in southeastern Australia, and in a large part of southeastern South America. Chicago's climate is a Df climate, with colder winters and hot summers; you'd experience something close to it in Moscow and Sapporo (northern Japan). The map shows you instantly where the "real" rainforest climate prevails, where the world's deserts are, and where the best grape-growing areas lie. Note that the Cs climate, a mild climate with a dry summer (which is denoted by the small letter s), occurs not only around the Mediterranean Sea (and thus in the famed wine countries of France, Italy, and Spain), but also in California, Chile, South Africa, and Australia. So you know what to expect, weatherwise, in Rome, San Francisco, Santiago, Cape Town, and Adelaide.

Of course it's possible to take part of this world map and draw it at a much larger scale. That takes away our ability to make worldwide

THE END OF THE ICE?

A few years ago I was invited to address the topic of climate change at the venerable Chautauqua Institution in New York, where every summer thousands of people of all ages gather to debate current issues, hear lectures, meet authors, attend opera and theatre performances and symphony concerts, and engage in a matchless range of intellectual pursuits. The daily lecture in the Amphitheatre attracts several thousand informed and critical listeners.

It was a hot July day, and people in the crowd were fanning themselves with their programs. It was not a good time to try to explain that while we do not know enough as yet about the periodicity of planetary climate change, we are now experiencing an ice age in which the current warm interglacial is only the latest of many. A twitter went through the audience and a few days later the local newspaper, *The Chautauquan*, published a humorous cartoon.

The following day a representative from the Pew Foundation spoke in the same venue, citing a huge array of statistics to confirm the obvious: the planet is warming. But then came a reference to "yesterday's speaker" as having been wrong: we are not in an ice age. I wondered whether the people at the Pew Foundation ever read *Science* or *Nature*. How easy it was to use another sweltering day to focus on anecdote and to obfuscate the long-range picture.

No doubt about it: the numerous cycles—axial, orbital, solar, atmospheric, oceanic—that generate nature's environmental seesaws continue even as humanity has become a major factor in the process during this interglacial through massive modification of the planetary atmosphere. But supercomputer models and IPCC projections notwithstanding, no one knows the proportional contribution to the current phase of climate change from natural and human sources. Contrary to what some well-intentioned scientists are asserting, we do *not* know with any satisfactory level of confidence what form climate change would be taking today in the absence of human interference. What *is* clear is that humans have become an additional factor driving climate change, and that reduc-

comparisons, but it increases the amount of information that can be provided for a smaller region. Travel kits almost never include such a detailed climate map. They should.

Ever since Köppen published his map of world climates, geographers and others have been intrigued by the apparent spatial relationship be-

ing the rate of pollution of the atmosphere should have priority as a public health as well as an environmental matter. But ice ages will continue to come and go. Glaciers will wax and wane. Sea levels will fall and rise. Species, cultures, and civilizations will flourish and fail. Nature's power will prevail.

tween certain climates and certain successful, powerful societies. Equatorial and tropical climates, apparently, do not favor the countries over which they prevail; none of the world's major powers, present or recent, lie here. Desert climates also do not appear to be conducive to big-time status. The same seems true of high-latitude, polar climates.

Are the mild midlatitude and the harsh continental climes the best our planet has to offer? Ellsworth Huntington had no doubt about it. "The people of the cyclonic regions," he wrote in 1942, meaning the midlatitude cyclonic zones, "rank so far above those of other parts of the world that they are the natural leaders . . . [they lead in terms of productivity, but] their greatest products are ideas and the institutions to which these give rise. The fundamental gift of the cyclonic regions is mental activity" (Huntington, 1942).

Such observations made Huntington famous; they also got him into a lot of trouble. By extension, it's possible to conclude that peoples living long term under cyclonic conditions would be intellectually superior to others, and from this, it was but a short step to nazi master-race philosophies. Environmental determinism, the idea that environment (chiefly climate) controls the capacities and destinies of societies, quickly got a bad name.

That was unfortunate, because Huntington had raised one of geography's central questions, and in the context of his time, he made enormous contributions to our understanding of climatic change and its impact on human societies. These days it is fashionable to join the chorus of disparagement, but many who do so have never read Huntington's last major (and monumental) work, *Mainsprings of Civilization*, which appeared in 1945, two years before his death. *Time* magazine, in its review of this book, proclaimed it comparable to one of the greatest of historical works, Arnold Toynbee's 12-volume *Study of History*.

Huntington argued that the seasonality of midlatitude climates had favored peoples who remained for many generations under these demanding, yet stimulating regimes. He explained the rise and fall of societies in terms of the "sweep" of climate change. Köppen's map, he frequently said in his lectures, should be seen as a still photo of a changing Earth—it represents the way things are today, but not as they were yesterday, nor as they will be tomorrow. When the Maya civilization rose to greatness, when West African states arose and prospered, when Muslim civilization flourished and out-shone its European contemporaries, it was because climate induced it. "The well-known contrast between the energetic people of the most progressive parts of the temperate world and the inert inhabitants of the tropics," he wrote, "is largely due to climate."

Small wonder that environmental determinism suffered the fate it did. Still, in recent years, the notion has been rediscovered and subjected to more rigorous analysis. Jared Diamond, who is on the faculty of the Department of Geography at UCLA, in his remarkable work *Guns, Germs, and Steel*, argues that it is not climate alone, but the environmental opportunities offered by a combination of natural conditions ranging from wildlife to plants and from water supply to terrain that put certain peoples at an advantage over others (Diamond, 1997). Lose those opportunities, and progress is halted. Benefit from them for a long time, and advantage endures.

To return to the theme with which this chapter started, we are living in a time of climate change to which, for the first time in our planet's history, a species, the human species, is contributing significantly. But make no mistake: the numerous interlocking cycles—solar, axial, orbital, oceanic—that generate nature's environmental seesaws continue even as humanity has become a major factor in the process through massive modification of the planetary atmosphere. But supercomputer models and IPCC (Intergovernmental Panel on Climate Change) projections notwithstanding, no one knows the proportional contribution to the current phase of "global warming" from natural and human sources. Contrary to what some scientists are asserting, we do *not* know with any satisfactory level of confidence what form climate change would be taking today in the absence of human interference. What *is* clear is that humans now play a role in driving climate change, and that reducing the rate of pollution of the atmosphere should have international priority as a public health as well as an environmental matter. But don't expect a reward in the form of "stopping climate change." Ice ages will continue to come and go. Glaciers will wax and wane. Sea levels will fall and rise. Species, cultures, and civilizations will flourish and fail. Nature's power will prevail.

6

THE GEOGRAPHY BEHIND WAR
AND TERROR

IT IS OFTEN SAID THAT "WAR TEACHES GEOGRAPHY," A PHRASE THAT emerged in the United States from the Second World War as the American public followed military campaigns in European and Pacific arenas of conflict and became familiar with geographic appellations such as Normandy and Iwo Jima. *National Geographic* maps hung on walls in military headquarters and in living rooms, battlefronts marked with red lines and advances with arrows as daily news reports chronicled the fortunes and misfortunes of combat. The agony of Arnhem and the horror of Hiroshima could not begin to be comprehended without the perspective of location.

The American public may never have been more geographically literate than it was after the end of that war. Geographers had played a role in the settlements drawn up at Versailles after the First World War, and they were to participate significantly in the postwar reconstruction planning following the Second. The national interest remained focused on struggling Europe, where a communist threat soon arose and where the Marshall Plan made a crucial difference. But the rehabilitation of Japan and continuing instability on the East Asian mainland also galvanized Americans' attention. In colleges and universities, geography courses covering Europe and Asia filled to capacity, and geography departments thrived. Not only was this a heyday of regional geography, educational precursor to the amorphous "area-studies" phenomenon that arose decades later; this also was a time when geographers' expertise was put to practical use. Shannon McCune was noted among these appointees: he served as Civil Administrator of Japan's Ryukyu Islands (of which Okinawa is a part) during the 1960s and was renowned for his insights into the problems of Korea, which he analyzed in an influential book (McCune, 1956).

As the communist threat faded in Western Europe, its revival on the Korean Peninsula plunged America and its allies into another war just a few years after the end of World War II hostilities, a conflict that seemed to presage a Soviet-American collision. Communist North Korea's armies in June 1950 invaded the South across the 38th parallel, the demarcation line agreed upon by Soviet and Allied forces to allocate the two parts of the peninsula to communist and Western control respectively. Flanked by a demilitarized zone (DMZ), the 38th parallel was not yet fortified the way it is today, and it provided no deterrent to the North Korean aggression. Before the ensuing three-year war ended with the loss of some 3 million lives, it had sown fear of communist expansionism, divided the American people (the war became increasingly unpopular and was criticized by 1952 presidential candidate Dwight Eisenhower), brought a threat by the new president to use nuclear weapons on North Korea and China, and set a precedent for American intervention to contain communist aggression.

Since the territorial theater of the Korean conflict was comparatively small, the dramatic seesaws of the conflict could be followed in large-scale detail on any wallmap. Americans became familiar with names such as Pusan (now Busan) on the southeast tip of the peninsula where four poorly-equipped American divisions hung on in a tiny bastion awaiting reinforcements; Inchon, where General Douglas MacArthur had his forces make the daring amphibious landing behind enemy lines that shattered the North Korean army; and the Yalu River marking the border between North Korea and China, toward which Allied forces were pushing the North Korean enemy when the Chinese regime decided to join the war. For more than three years the Korean War and its associated dramas riveted, involved, and dismayed the American public. Indeed, the public got engaged more than ever when MacArthur tried to appeal directly to the American people after being rebuffed by President Truman when the General wanted to blockade China's coast and bomb Chinese military bases.

Undoubtedly the Cold War and the specter of a potential Third World War helped sustain American interest political geography. During the later 1950s, as the world became a stage for ideological competition and proxy wars, the Soviet Union became the next focus for geographic analysis. A growing number of articles in the geographic literature, and an increasing number of *National Geographic* map supplements, reflected this preoccupation. If the colonial empire that was the Soviet Union never got imprinted as clearly on Americans' mental maps as was Europe during and after the Second World War, this was no surprise: an empire extending from the Iron Curtain across Europe to the shores of the Pacific was a world unto itself. Subsequent miscalculations about the USSR's ability to feed itself, Moscow's military capacities and estimates of the Soviet Union's capability for survival proved that it was not just the public whose images of the place were weak.

AMERICA IN INDOCHINA

But the next war waged by the United States to contain communist expansionism was the one that added a word to the phrase about geography teaching war . . . *belatedly.*

In truth, Southeast Asia had not been an American priority even during World War II ally France's struggle to re-establish its dominance in Vietnam prior to its infamous defeat at Dien Bien Phu in 1954. Neither did Britain's difficult but ultimately successful suppression of a communist insurgency on the Malayan Peninsula draw much attention, but of course both campaigns occurred during the first half of a decade dominated by the costly Korean War. Vietnam took center stage when the geopolitical implications became clear of a communist regime centered in North Vietnam's capital of Hanoi and backed by the Soviet Union and China, with a U.S.-supported regime governing from South Vietnam's capital of Saigon. This was the result of the negotiated settlement following the French defeat, creating a boundary between the two countries along the 17th parallel flanked, as in Korea, by a DMZ.

Unlike the Korean situation, this division of Vietnam was to be followed, according to the settlement signed in Geneva, by free elections to be held in 1956 throughout both Vietnams in order to install a single, democratically elected government. The North Vietnamese under Ho Chi Minh, who had not only led the anti-colonial campaign against the French but had also built a broad and popular political organization north as well as south of the DMZ, were expected to win this contest, but Ngo Dinh Diem, who had seized control of the South, refused (with American approval and support) to hold the scheduled vote. This motivated the North Vietnamese to try to unify Vietnam through military force, and a twenty-year war (1956 to 1976) ensued.

How much of this sequence of events was seen in geographic context by those on this side of the Pacific who were prosecuting the war or those affected by it? The two decades of war had a wrenching impact on American society, exposed civilian leaders as liars, divided the society, coarsened the culture. Optimistic forecasts by military leaders were contradicted by evidence from the field brought into living rooms by media reports. Civil disobedience disrupted streets and classrooms. When American and South Vietnamese forces expanded the "Vietnam War" into eastern Cambodia in 1970, U.S. aircraft began bombing northern Laos, and the Vietnam War became the Indochina War, American citizens responded with a wave of anti-war demonstrations and protests. Universities and colleges were the flashpoints of opposition.

But when the war ended in defeat and journalists and educators in America randomly polled citizens and students for their reactions, it was clear that while political opinions were strong, mental maps were not. One Midwestern college, asking incoming first-year students to identify Vietnam on an outline map of the world, reported that only 7 percent (1 in 14) got it right.

Not until decades later did it become clear that insufficient knowledge of the physical- and cultural- geographic complexities of Vietnam in particular and Indochina in general afflicted not just opinionated citizens, but also, more disastrously, leaders, elected and appointed, who propelled the country into this calamitous conflict in the name of communist containment. Robert McNamara, U.S. Secretary of Defense from 1961 to 1968 and who played a key role in the campaign, brought back optimistic predictions of impending North Vietnamese failure, and became President Lyndon Johnson's chief deputy guiding the conduct of the war. Although he began to express doubts late during his tenure and broke with President Johnson over the issue of continued carpet-bombing of the North, it was not until long afterward that he acknowledged his failure as a guide to American policy and strategy (McNamara, 1995).

In this retrospective, McNamara lists "eleven major causes for our disaster in Vietnam," including the following: "Our misjudgment of friend and foe alike reflected *our profound ignorance of the history, culture, and politics of the people in the area* (italics mine). We underestimated the power of nationalism to motivate a people . . . and we continue to do so in many parts of the world."

Nothing else in McNamara's list even approaches the importance of these two sentences. Exaggerating the dangers posed "by North Vietnam and the Vietcong to the United States," while foolish, hardly matches such profound geographic ignorance, nor does "(our) failure to recognize the limitations of modern, high-technology military equipment, forces and doctrine" or "we had not prepared the (American) public to understand the complex events we faced." Nor is (our failure to) "recognize that neither our people nor our leaders are omniscient" a comparable cause. Indeed, the majority of McNamara's "causes" are internal to the United States: bureaucratic failures and didactic shortcomings.

It is noteworthy that the McNamara retrospective on Vietnam includes not a single map of Indochina's intricate ethnic diversity. Two endpaper maps, identical in the front and back of the book, show a (very) rough outline of the states of Southeast Asia and a detail of Indochina that contains the "Corps Area Boundaries" the sectors of South Vietnam assigned to U.S. Army Corps, and the Ho Chi Minh Trail. No map in the book shows the domains of the minority Hmong, Muong, Tho, Thai, Khmer, Nung, Hoa, or Dao, although disaffected minorities played a disproportionately large role during the war. A map of religious affiliation before and after partition would have been enlightening given the role of Buddhists, Roman Catholics, "New" Religionists, and Traditionalists in the struggle. Apparently the "lessons of Vietnam" posited in the book's subtitle were not fully absorbed.

Secretary McNamara (M.B.A., Harvard University) would have benefited from a graduate course in cultural and regional geography that might have sensitized him to such issues as moral codes and value systems, ethnic links and family ties, religious codes and language patterns. Unfortunately, not to say disastrously, Harvard University is not sufficiently universal to offer undergraduate or graduate geography to the future leaders who enroll in its classes.

CURRICULAR CASUALTY

Compared to the enormity of the economic and social costs of the Indochina War, what was simultaneously happening to America's educational system may not loom large in retrospect. But it was crucial nonetheless. The war strained the fabric of society, creating generational mistrust, suspicion of authority, negation of cultural standards, breakdowns of civility. Disruption of university and college classes, "teach-ins," forcible occupation of faculty offices and administration buildings, even the forced confinement of presidents and provosts by armed demonstrators made the 1960s a turbulent time for higher education. Such incidents made the news, but order was restored surprisingly soon after the conflict ended.

Not all of this was the direct result at home of the Vietnam conflict abroad. The United States was in the aftermath of the civil-rights movement's apogee, which by itself would have posed a major challenge, not least to schools and colleges. But even as President Johnson's "Great Society" foundered on the rocks of war, normalcy returned to America's campuses—or so it seemed.

Behind the scenes, changes in American education—at all levels—proved to be far more durable and less reversible. University core curriculums were dismantled. Grade inflation, especially in the social sciences and humanities, soared. "Pass-Fail" options replaced measured achievement. "Life experience" was deemed by some educators to merit the award of academic credit. K through 12 education underwent wrenching changes, many of which, the "New Math" among them, turned into expensive failures. As noted in Chapter 1, geography lost its identity as a discrete school subject as education's innovators amalgamated subject matter and erased topical boundaries. And since geography, among these subjects, was the only one to straddle the natural and social sciences, the "social studies" product of this reorganization deprived millions of students of an early introduction to Earth science as well as a first look at the way nature and humanity interact. That these interactions are most effectively studied using maps was another casualty of the process.

Geographers of every rank warned of the consequences, pointing out that growing geographic illiteracy would come to constitute an even greater liability than it had been during the Indochina War. McNamara in his book complained that "We failed to retain popular support in part because we did not explain fully what was happening and why we were doing what we did. We had not prepared the public to understand the complex events we faced and how to react constructively to the need for changes in course as the nation confronted uncharted seas and an alien environment." But his and his cohorts' acknowledged and "profound ignorance of the history, culture, and politics of the people in the area" would have invalidated such explanations in any case. Whether the public understanding involves climate change or foreign wars, there is nothing like a strong foundation in physical and human geography to underpin that comprehension. And that goes for the people as well as the people's elected representatives.

A WORLD IN TRANSITION

The containment-of-communism policy that had guided American and Allied actions in Indochina and Korea gradually devolved, during the period of decolonization, into overt as well as covert participation in proxy wars, conflicts and struggles in countries, many of them newly independent, between pro-Western and pro-Soviet interests. To be sure, the period of decolonization greatly complicated the world map. New names appeared, from Bangladesh to Burkina Faso and from Sri Lanka to Zimbabwe. Some came and went (East Pakistan, Zaïre). Some of the new (and old) countries in the global periphery now became ideological battlegrounds, for example Afghanistan and Ethiopia. While the decolonization period lasted for several decades and changed the world map (and the United Nations membership roster), there was little hint of the even more momentous transformation that lay ahead.

By the time the United States Government once again sent a major force into combat, the theater of war was not in Africa or South Asia, but in the Middle East, 15 years after the end of the Indochina War. In what was to become known as the First Gulf War, American, European, and Arab forces in January 1991 launched a campaign to oust the Iraqi army from neighboring Kuwait, which Saddam Hussein had invaded nearly six months earlier. In six weeks, an Allied coalition of about 700,000 troops ousted 300,000 Iraqis and restored Kuwait's independence after Saddam Hussein had declared it to constitute Iraq's "19th province." It was a clearly-targeted action notable for the remarkable cross-cultural alliance assembled by President George H. W. Bush, for its clear geographic target, and for the restraint displayed by the victors following Iraq's ouster from Kuwait. Although a large part of Iraq's Republican Guard was destroyed and the way to Baghdad appeared open, no invasion or overthrow of the Iraqi regime followed.

While this brief but violent campaign focused world attention on a volatile area, the real focus of global transformation was not (yet) in the Islamic realm, but in the Soviet Union and Eastern Europe. The collapse in December 1991 of the world's largest state, the U.S.S.R., ended a Cold War that had shadowed the planet for nearly half a century. Out of this communist empire arose 15 independent republics whose names and locations, except for Russia itself, were anything but familiar on most Americans' mental maps. The same year witnessed the start of the violent disintegration of Yugoslavia, complicating the map of Eastern Europe as no fewer than seven small countries emerged from the wreckage (the most recent, Kosovo, still was not universally recognized as a sovereign state in 2012).

It is not surprising that some scholars saw in this harrowing scenario the approaching end of the state as we have known it, the last gasp of true sovereignty and the start of a "new world order." States, they reasoned, would not be able to get along without each other's support in this fast-changing world; already, European states were joining in a European Union that might be a precursor to a United States of Europe. The United States was part of NAFTA; Russia was trying to resurrect a union of former Soviet republics. I referred to this hypothesis during a 1999 lecture at Rice University's Baker Center attended by Secretary of State James Baker himself—who promptly and forcefully reproached me for bringing it up. "This may be a time of turmoil," he said, "but mark my words. The state will be the dominant player in international affairs for a very long time to come."

It may be reasonable to assume that if Americans' mental maps were clearest right after the Second World War, they were never more vague than at the turn of the century. The bewildering transformation of the world's political geography tested even the most attentive observer. But even as Russia teetered on the verge of instability and casualties mounted in Yugoslavia's collapse, two other related developments, up to that point not as dramatic, had the attention of geographers and others trying to anticipate what lay ahead. The first had to do with the global picture: the kind of "new world order" that seemed to be arising, and how it would emerge. The second was a series of actions by terrorists espousing various causes, ranging from airplane hijackings to deadly assaults on military, diplomatic, and civilian targets.

GLOBAL CIVILIZATIONS AND GEOGRAPHIC REALMS

For a half century political geographers have speculated on the topic of conflict and its manifestations in the form of clan warfare and tribal strife, feudal contests and interstate hostilities, two world wars and the potential for "civilizational" clashes arising from a combination of ideology,

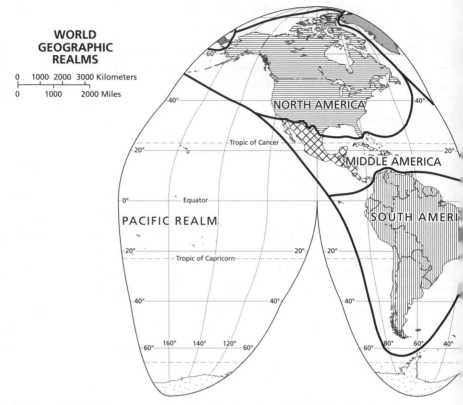

Fig. 6-1. The world divided into geographic realms. Not shown here is the division of each realm into component regions, for example West Africa, East Africa, Equatorial/Central Africa, and Southern Africa.

religion, territorial ambitions, and economic interests. Figure 6-1 appeared in introductory geography textbooks as early as the 1970s and has been a standard topic in political geography for even longer (de Blij, 1971).

It is worth having a close look at this map, because it displays the world of the twenty-first century as a dozen "geographic realms," of which eight are dominated by a large and actually or potentially powerful state and three contiguous ones are marked by unifying histories, belief systems, or ethnicity. These three (Europe, Subsaharan Africa, and North Africa and Southwest Asia) also are by far the most fractured politically.

Note that the United States, Mexico, Brazil, Russia, China, India, Indonesia and Australia are the dominant states in their respective realms,

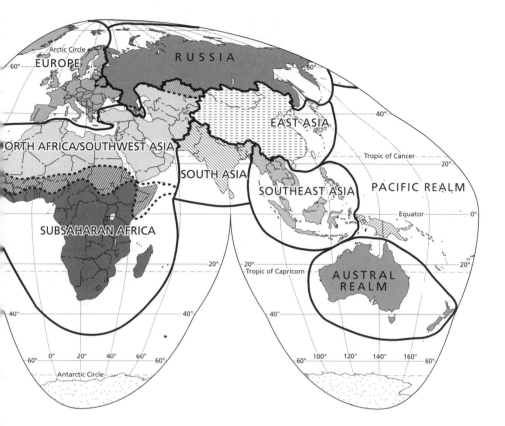

Indonesia having least lived up to its potential as the leading state in its realm.

Is this map a guide to "civilizational" clashes in the future? A comparison to a map on pp. 26–27 in a much-debated book on this prospect would make it seem so (Huntington, 1996). The argument that the world geopolitical stage is being reconfigured along cultural fronts, and that civilizational conflict will erupt, in Huntington's parlance, as cultural "fault line wars" was not new in the 1990s, at least not in geography. (The allusion to geomorphology is rather unfortunate inasmuch as a fault line is an erosional feature rather than the real thing). But Huntington's hypothesis, which holds that the remaking of the world order will essentially pit the collective "West Against the Rest," started a discussion that still continues. The realms of intuitive consensus of the 1960s have become far more clearly bounded, yielding culturally discrete entities whose internal cohesion is partially based on growing antipathy to the West. The West of the 1960s, meanwhile, has itself become quadripolar, with the (comparatively) waning superpower United States in an increasingly uneasy relationship with a wary Europe and with tendentious ties to an unpredictable Russia. This, even as Brazil and greater South America loom larger on the global stage. Huntington predicted that the ultimate "civilizational clash" would pit the mainly Christian West against the dominantly Islamic realm, and marshaled substantial evidence in support of that notion, but the potential for another contest involving China may be greater (see Chapter 8).

Nevertheless, the backdrop events of the 1990s did seem to presage a cultural conflict of larger dimensions involving the Islamic realm and its neighbors near and far. In response to criticism of his assertion Huntington enumerated all the "active" conflicts and violent incidents of the decade, and concluded that more than half of them involved Islamic sources, either in the form of Islamic versus non-Islamic targets or intra-Islamic strife such as Sunni against Shia (or other sectarian) hostility. In the Western world these actions seemed to represent a form of warfare that, while not new and certainly not unique to Islamic practitioners, was unsettling, its geography frighteningly random, and its objectives often unclear and not negotiable.

TERRORISM'S WIDENING CIRCLE

Successful acts of terror have a way of imprinting themselves on the culture of their victims. On the morning of May 14, 1940, four days after nazi armies had crossed the Dutch border on their way to the shores of the North Sea, the skies over the port city of Rotterdam were suddenly filled with warplanes. Without warning they rained incendiary bombs on the city's historic center and the harbor. Hundreds were killed in the first wave of the assault, and when the survivors tried to douse the flames with fire hoses they were shot by nazi paratroops who had landed at the airport.

The attack was a signal to the Netherlands' government, whose forces were having some unexpected success in resisting the German aggression, to capitulate or risk further death and destruction.

Was this an act of terror or an act of war? And if the former, was Dresden any different? That German city was targeted by Allied bombers beginning in mid-February 1945 for reasons similar to Rotterdam in 1940: to persuade a government to capitulate and thus shorten the war. Perhaps as many as 100,000 civilians were killed, but the sustained campaign had no effect on nazi policy. How about Hiroshima and Nagasaki? The argument over casualties continues: would more American and Japanese soldiers (as well as civilians) have died in the inevitable invasion than perished in the two cities destroyed by United States atomic bombs? Is there such a thing as state terrorism, and if so, do these four cities exemplify it? Certainly the Palestinians have no doubt about the reality of state terrorism. The Palestinian press regularly accuses Israel of it when Israeli warplanes and tanks in pursuit of Arab terrorists kill civilians in Gaza or the West Bank.

The definition of terrorism is no minor issue. It raises heated arguments in the scholarly literature, and it has important legal and fiscal (notably insurance) ramifications. In a book entitled *The Geographical Dimensions of Terrorism*, terrorism is defined as "intimidation through violence," but without further qualification (Cutter et al., 2003). Neither is the definition written by the United Nations Panel on Threats, Challenges and Change satisfactory: Terrorism is "any action intended to kill or seriously harm civilians or non-combatants, with the purpose of intimidating a population or compelling action by a government or international organization" (Annan, 2004). A more considered view defines terrorism as "an unprovoked attack against civilian noncombatants away from any theatre of war by men or women not working for a power openly at war with the victims' country" (Harvey, 2003). The problem with such carefully qualified definitions is that further definitional caveats arise: what is "unprovoked?" Terrorists claim a host of provocations, from cohorts' deaths to historic injustices. Does "openly at war" mean that a rebel movement's declaration of war on a government— which has occurred repeatedly—is enough to negate that clause and thus the charge of terrorism? To such questions, there is no simple answer.

A RISING TIDE OF GLOBAL TERROR

"One person's terrorist is another person's freedom fighter" goes a common refrain when it comes to violence to achieve political goals. When the Irish Republican Army (IRA) claimed responsibility for a bomb attack on Prime Minister Margaret Thatcher at the Grand Hotel in Brighton in October 1984, killing 5 and wounding 31, British public opinion held this to be an act of terrorism whereas many Catholics in Northern Ireland viewed it as justified resistance. When a Palestinian suicide bomber destroys a bus full of civilians on an Israeli street, Jews see an act of terror, many Palestinians a heroic strike at a hated enemy.

In these and many other examples known to all of us, the violence is an outgrowth of national policies. The IRA fought against British rule in Northern Ireland (the issue is closer to being settled today, but no final resolution has been achieved and the potential for further upheaval remains). Palestinians fight against an Israel they resent as a trespasser, invader, and occupier. Corsicans in France, Basques in Spain, Tamils in Sri Lanka, Chechnyans in Russia, Uyghurs in China—in every instance, their battle is or was against the state that dominates them, and in every case they carried their violent campaigns to the major cities of their adversaries. Corsicans bombed the city hall of Bordeaux; Basques assassinated politicians and civilians from Madrid to Barcelona; Tamils killed hundreds in Colombo, Chechnyans turned Moscow into a war zone, Uyghurs detonated bombs in Urümqi. None of the longstanding issues provoking these actions is yet resolved except perhaps the Tamil campaign, but their protagonists are well known and their geographies are tangible. What the activists want is a redrawing of the map: an autonomous Corsica, an independent Basque territory, a partitioned Sri Lanka, a sovereign Chechnya, a free East Turkestan.

No such geographic clarity marks the form of terrorism that has become dominant in the twenty-first-century world: Islam-inspired violence. A generation ago, acts of terror punctuated political strife in numerous places around the world, affecting not only the locales where it was endemic but also such countries as Canada, Germany, Italy, India, and Japan. In the United States, occasional terrorist acts tended to be perpetrated by far-right or neo-nazi activists, often operating individually. In Norway in 2011, a deranged extremist vented his anti-immigrant views by killing nearly 80 people in a shooting rampage in and near the capital, Oslo. Elsewhere, shadowy groups with diverse grievances and objectives used bombs, guns, hijackings, kidnappings, and other means in pursuit of their objectives, which often seemed to be, simply, to create mayhem. Until the early 1980s, although some cooperation did occur, there was no global network coordinating these groups. But circumstances were evolving. Terrorism was becoming more than a local crime problem to be dealt with by local law enforcement. A new and larger wave of terrorist activity was rising, overshadowing all the local movements and creating a global threat, targeting Western interests worldwide. This terrorist network had religious rather than ideological foundations and exhorted Muslims everywhere to participate. It found a ready market among Salafists (Salafism demands strict emulation of Islam's seventh-century founders) and others who espoused Islamic revivalism in its most fundamental forms.

Even so, little was known about this fundamentalist movement just three decades ago. Islamic terrorism had for some time been equated with Palestinian Liberation Organization (PLO) activity against Israel and Israeli interests, ranging from the deadly attack against Israeli athletes during the 1972 Olympics in Munich to the 1985 hijacking of an American airliner leaving Athens by a group of Shi'ite Muslim terrorists demanding the release of fellow Shi'ites in Israeli prisons. The widening circle of this violence sometimes struck United States interests, but its source remained obscure. In 1982, the name "Islamic Jihad" (Islamic Holy War) began to appear on intelligence reports, a shadowy organization that claimed responsibility for four terrorist attacks during 1983 and 1984 on American military and diplomatic installations that killed nearly 320 people. Yet, as one perceptive observer wrote at the time, "there [are] doubts that Islamic Jihad, as a proper organization, actually exists. No names or locations have been publicly identified . . . it is widely believed that 'Islamic Jihad' is a convenient cover name for a host of fundamentalist movements and cells throughout the Middle East" (Wright, 1986). Speculation as to Islamic Jihad's headquarters centered on Shi'ite Iran, in part because of Lebanon-based Hizbullah and its known ties to Tehran. But in truth there was little concrete information about an enemy proving itself capable of inflicting heavy casualties on American and other Western interests.

SOUTH ASIA AND SAUDI ARABIA

Even as such attacks continued, the seeds for a far more serious challenge to regional security were being sown elsewhere. After the Soviet Union invaded overwhelmingly (Sunni) Muslim, landlocked Afghanistan in 1979 with the aim to establish a secular regime in Kabul, the United States made the fateful foreign-policy decision to support anti-Soviet *mujahideen* (strugglers) by providing them with large quantities of modern weapons as well as money. It was a Cold War strategy that might be charitably described as shortsighted, and certainly appeared contradictory: a United States supposedly committed to principles of secular government helped oust a regime attempting to establish just that, albeit undemocratically. Soviet restraint in the United States struggle for Vietnam was not rewarded with American moderation in Afghanistan, even when it must have been obvious to some policy makers in Washington that a forced withdrawal of Soviet troops would leave Afghanistan a destabilized, faction-ridden, failed state flooded with armaments at a time when Islamic terrorism was on the rise. Soon the factions that had been somewhat united during the anti-Soviet campaign were in conflict, with warlords controlling fiefdoms far from the devastated capital and more than 3 million refugees encamped on the Pakistan side of the border.

The United States, of course, was not the only factor in the struggle for Afghanistan. Another was Saudi Arabia, source of megafunds in support of the mujahideen and wellspring of Islamic extremism that would find fertile soil amid the cultural chaos that followed the Soviet defeat. For decades, Saudi money had supported numerous religious schools or *madrassas* in Pakistan, schools in which education is confined to intensive study and memorization of the Quran (Koran). These schools had their origins in British colonial times, when Muslims found themselves not only colonized and deprived of their power but also a minority in the dominantly Hindu South Asian realm. In response, as early as the 1860s, Sunni Muslim communities set up such deobandi schools. The cleric-teachers of these madrassas were able to issue *fatwas* (legal, interpretive proclamations based on the holy texts of Islam, the Quran or the Hadith, that stipulate how Muslims should live their lives, and thus to protect Muslim communities from undesirable influences). When Pakistan was severed from India and became an Islamic state, the Pakistani madrassa network became a potent factor in national life. Supported by taxes and drawing youngsters by the hundreds of thousands from the Sunni majority (the Shi'ite minority generally disapproved and objected to being taxed for this system), the madrassas provided lodging, food, and religious education mainly for the sons of poor families. During the war in Afghanistan, when several million Afghans took refuge in Pakistan, many of their sons also enrolled in these schools. Years of indoctrination and rote memorization left them poorly prepared for the real world, and the religious perspective with which they graduated was—and is—retrograde.

In the eyes of their Saudi supporters, however, there was—and is—nothing retrograde about Pakistan's madrassas. Indeed, their teachings tend to conform to those of fundamentalist and often extremist Wahhabi dogma that pervades conservative Saudi religious proselytism. The Arabian-born theologian Muhammad ibn Abd al-Wahhab (1703–1792), the founder of this movement, was determined to bring Islamic society on the peninsula back to its most traditional and puritanical form. So strict were his teachings in his hometown near Medina that he was expelled by the locals in 1744, when he moved to what was then the capital of the central territory of Najd, whose ruler was named Ibn Saud (that old capital lies close to the modern one, Riyadh). That was a fateful move, because Ibn Saud liked al-Wahhab's teachings and formed an alliance with him that was facilitated by the fortune al-Wahhab had inherited upon his wife's death. Together they set out on a path of conquest, political and religious, that created a ruling Saudi dynasty over most of the Arabian Peninsula and made Wahhabism the dominant dogma. In 1932, when the Kingdom of Saudi Arabia was officially established, the ruling king made Wahhabism the state religion.

The era of oil production, foreign corporations, modernization, and immigration by millions of Arabs and non-Arabs from sources ranging from Palestine to Pakistan changed Saudi Arabia substantially, but Wahhabism survived. The Saudi royal family had grown by the turn of the century to nearly 5,000 members, and while some had Western educations and others had foreign experience in diplomatic and business fields, still others followed Ibn Saud's traditions and remained staunchly conservative, supporting fundamentalist religious institutions within Saudi Arabia and mosques and schools abroad. In Pakistan's madrassas these conservatives saw the values they embraced, and in conservative clerics they found allies in support of their revivalist cause (many Muslims prefer the term "revivalist" to "fundamentalist," arguing that the former is appropriately forward looking). Thus money flowed to mosques and schools abroad, fueling Islamic revivalism while creating channels to funnel such money to fanatics willing to take up arms, and give their lives, on behalf of extremist causes.

From Pakistan came the products of the madrassas, from Saudi Arabia came money. But there was a third element in the transformation of Afghanistan from isolated backwater to terrorist base. In several ways, the life of Usama bin Laden mirrored that of al-Wahhab: he gained significant wealth through inheritance, in his case from his father; following college graduation he left Saudi Arabia to pursue his causes elsewhere; he returned to denounce his leaders' lack of commitment to Islamic principles; and he was exiled—not just from his local community, but from his country. During the 1980s, though, Usama bin Laden and the United States had a common goal: to oust the Soviets from Afghanistan. He used his own money and an even greater hoard from Saudi royal and religious donors to propel this campaign, watching his fame as well as his war chest grow in the process. In 1988, even before the last Soviet troops had left Afghanistan, he and a group of associates founded an organization called al-Qaeda. Following the final withdrawal of Soviet forces, he returned to Saudi Arabia, leaving behind a large network of allies in the region. Once back in his home country, he publicly denounced his government for allowing United States troops on Saudi soil during the 1991 Gulf War, when an alliance of countries including Arab states, led by American forces, ousted the Iraqi army from Kuwait. By this time bin Laden had become a prominent figure, and the Saudi regime wasted no time in again expelling him from the country, this time stripping him of his citizenship. Allegedly with the help of a conservative member of the royal family, he found refuge in Arab-ruled Sudan, where he established a number of legitimate businesses to facilitate his now-global financial transactions, but also set up several terrorist training camps. Next he convened conferences attended by terrorist leaders and activists representing an ever-growing number of allied organizations ranging from Jemaah Islamiya (Indonesia) and Abu Sayyaf (Philippines) to Islamic Jihad (Egypt) and Hizbullah (Lebanon). When this became known, the Khartoum regime came under strong international pressure to oust him, and in 1996 bin Laden and his circle of confidants transferred their headquarters back to Afghanistan.

By now bin Laden's role as leader of a terrorist network far more extensive than Islamic Jihad was well understood, and his organization, al-Qaeda, had claimed responsibility for numerous major terrorist strikes, including the first attack (1993) on the World Trade Center in New York City killing six and injuring about one thousand people. Indeed, bin Laden issued a series of *fatwas* against the West and the United States, in one of them actually declaring "war" on America. Bin Laden's forced relocation placed him in physical as well as cultural settings that yielded crucial advantages. The cave-riddled, mountainous terrain of eastern Afghanistan offered numerous hideouts, with limited and controllable accessibility (bin Laden's taped, open-air threats and exhortations, broadcast on Western television, actually induced some geologists to try to identify rock formations to help counter-terrorism specialists to identify his whereabouts). The tribal and clan-ridden cultural landscape of the area enabled bin Laden to use his wealth and influence to protect his seclusion; Mullah Omar, the Afghan Taliban leader, became his most effective ally. It was a combination of geographies that gave the al-Qaeda kingpin a hideout in which to plot his strategy and a perch from which to stoke the anger that would energize his operatives.

GEOGRAPHY OF RAGE

On the walls of many a geography teacher in the Muslim realm hangs a map not unlike Figure 6-2, depicting all areas of the world that are, or were at one time, under Islamic sway. That map shows a contiguous Muslim *umma*, or world, extending from West Africa to Central Asia and from Eastern Europe to Bangladesh, with an outlier in Southeast Asia. It includes most of Spain and Portugal as well as Hungary, Romania, Bulgaria, and Greece, much of India and part of western China. It prompts memories not only of Islam's global reach but also of its early splendor as a culture whose achievements in science and mathematics, architecture and the arts far exceeded those of Europe. Refer to this map, and even "moderate" Muslims tend to become quite agitated. "I remind myself daily of the humiliations we have suffered and the lands and peoples of God we have lost," a teacher at an otherwise modern, well-equipped school in Alexandria, Egypt told me in 1980. "This map is our inspiration, our minimum demand on the world." I was unaware at the time of the portent of those sentiments, versions of which I had already heard in South Asia and East Africa and would hear again in Dubai and Morocco. I thought of this map when Usama bin Laden, in one of his post-9/11 taped diatribes, referred to the "loss" of al-Andalus, now part of Spain, among his justifications for the carnage he had caused.

If ever there was a "civilizational" mental map, Figure 6-1 is it. While this image may be more firmly planted in the Arab world than in, say, Bangladesh or Indonesia, it is seen in Muslim schools from Morocco to Malaysia and from Kosovo to Kenya. It is Islam's geography factor, although it generalizes a very complex cultural mosaic. It depicts a Muslim world idealized across time as well as space: at no single time did all the areas shown as Islamic fall under Islam's sway. It is a source of pride, but also dishonor. Such a sense of shame and mortification is widespread, if not universal, in the contiguous Muslim world: Muslims once ruled an Ottoman Empire that reached from present-day Turkey to the gates of Vienna and a Mogul Empire that extended from modern

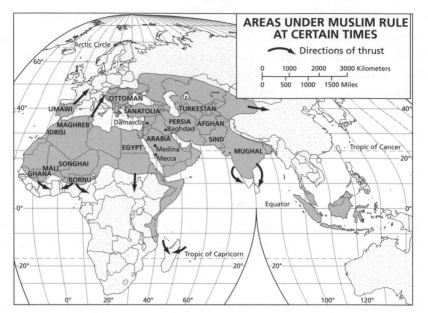

Fig. 6-2. The domain where Muslims ruled has expanded and contracted over time (for example, Muslim rule was ended in Iberia and later in Eastern and Mediterranean Europe). But many Muslims believe that any area once under Islam must be recovered. This map approximates what this would require.

Pakistan to Bangladesh. They lost Iberia and they were ousted from almost all of Eastern Europe. They were battered by the Crusaders and colonized by the Europeans and the Russians, whose geometric boundaries of administrative convenience created eventual states of impractical configuration (Iraq was one of those states). They had no voice when, in the aftermath of a war among Western powers, the United Nations carved out a state of Israel in their Middle Eastern midst. And then the West's voracious appetite for oil brought foreign economic, cultural, and political penetration of what remained of the Islamic realm, even of the holiest of lands, the Arabian Peninsula, site of Mecca and Medina.

Thus there is no reason to ask, in the aftermath of 9/11, what it is Westerners in general or Americans in particular have recently done to persuade 15 middle-class Saudis and four other Muslims to send more than 3,000 people to a fiery death in coordinated suicide attacks so terrifying that they unhinged the civilized world. In the annals of history, what was done to Islam and Muslim peoples is proportionately no more dreadful than what was done by European (and Arab) enslavers to Africans, American settlers to Native Americans, Belgians to Congolese, Germans to Jews, and too many other depredations to chronicle. But Africans are not engaged in suicide missions in Brazil, Native Americans are not bombing United States cities, Congolese are not targeting Brussels, and Israelis are not blowing up German commuter trains. If settling historic scores were to have become routine, the planet would no longer be a livable place.

It is often argued that the answer to the question just posed lies in the Israeli-Palestinian struggle and the consequent radicalization of Islamists everywhere, creating a growing cadre of terrorists ready to give their lives in the cause of Allah. But there is little evidence for this. The cycle of violence that has ensnared Israel and the Palestinians has more in common with Northern Ireland, the Basque problem, and Tamil Eelam than with al-Qaeda's campaign, which means that even a two-state solution to the satisfaction of both parties will have little or no impact on the larger issue. No Palestinians are known to have been involved in the planning, funding, or execution of the 9/11 attacks. A majority of Palestinians, their conflict with Israel notwithstanding, would favor a peaceful and territorially fair resolution and would accept Israel as their neighbor. This is not the position of al-Qaeda or its allies, and it is not what that map of the immutable Muslim world reflects.

No amount of behavior modification on the part of the Western world or of the United States can undo what history has wrought, and of course there is no way to return the planet to the circumstances represented by Figure 6-2. Not even a complete cessation of oil exports and the total withdrawal of all Westerners and Western interests from Saudi Arabia would have been enough to satisfy bin Laden and his Wahhabist associates: they equated the "moderate" wing of the royal family with the former shah of Iran, and nothing short of a theocracy of the Khomeini variety would do. Indeed, Khomeini himself made a move that reveals the intent of those who espouse the true faith: in 1989 he issued a fatwa that reached beyond the world of Islam, the umma, by proclaiming a death sentence against a British author living in the United Kingdom for a work allegedly containing blasphemy. This pronouncement compelled Muslims to attempt to find and kill the offender, who had to go into hiding in his own country. Not just the Islamic world, but the entire world must countenance the laws of Islam.

Let us look again at Figure 6-2 and compare it to maps of world natural environments, especially climate (Fig. 5-3). It is instructive to note that the contiguous region where Islam prevails today (thus the area shown minus Iberia and most of Eastern Europe; Islam no longer rules but still has strength in India) coincides remarkably with the world's harshest desert climates. Indeed, the harshest forms of Islam seem to prosper in the toughest environments: Saudi Arabia, Pakistan, Sudan; the milder forms of it appear to predominate under more moderate environments in Indonesia, Malaysia, and Bangladesh. This is not to suggest a causative relationship, but it relates to another observation of the spatial character of the faith, which is that its core area remains most fundamentalist while the periphery (not only Indonesia but also Turkey, Morocco, and Senegal) tends to be more temperate. This may be taken to a larger scale: on the Arabian Peninsula itself, coastal Dubai, where women (unlike Saudi Arabia) may drive, and Oman, where some schools admit girls as well as boys, conditions in Islamic society vary. In the Middle East, Saddam's Iraq was a secular state under the Sunnis' harsh rule; its leaders wore business suits or military uniforms, not religious garb. In Lebanon, about one-third of the Arab population remains Christian (down from 50 percent a half century ago), and even some Palestinians remain Christians today. In the Maghreb, the Atlas Mountains form a cultural divide as well as a physical one: the coastal zone with its more cosmopolitan cities and towns and their busy bazaars are a far cry from the Berber interior and its villages and caravans.

Does this mean that moderation has a chance, that urbanization, migration, globalization, and economic interaction will eventually temper the rage that accompanies the revivalism driving Saudi Arabia's angry conservative clerics, their fiscal sponsors at home and their Wahhabist allies abroad? It does not appear so. In the aftermath of 9/11, well-intentioned American colleges began teaching courses about Islam and the Quran, but Islam's holy book, like the Bible, makes discouraging and well as difficult reading. True, it contains hopeful sentences such as "There shall be no compulsion in religion" (Quran, 2:256), but anyone hoping that the contradictions familiar to Bible readers may be fewer in this holy book will be disappointed. "On almost every page, the Koran instructs observant Muslims to despise nonbelievers. On almost every page, it prepares the ground for religious conflict" (Harris, 2004). That bleak assessment actually understates the case. The Quran's angry denunciations of all who qualify as "infidels," its warnings against interaction with "unbelievers" and its promises that "those that deny Our revelations will burn in fire . . . no sooner will their skins be consumed than We shall give them other skins, so that they may truly taste the scourge"

(Quran, 4:55) are read by the faithful and elaborated by mullahs in sermons too rarely marked by restraint. In Islam, Harris argues, "the basic thrust of the doctrine is undeniable: convert, subjugate, or kill unbelievers; kill apostates; and conquer the world."

Thus the territorial imperative reflected by Figure 6-2 is in fact exceeded by the Quran's stipulations, and in addition it is matched by the faith's position toward those who might wish to change religions from Islam to, say, Buddhism, Hinduism, or Christianity. Numerous Christians have converted to Buddhism; others have converted to Islam, some now-prominent Americans (found fighting for Islamic causes) among them. But in Islamic law—and, among the major religions, Islamic law alone—conversion from Islam is apostasy, punishable by death. Once a Muslim, always a Muslim; attempt to renounce the faith, and both the convert and he who encouraged the conversion are condemned. The Hadith is quite specific on this point: "Whoever changes his religion, kill him" (Hadith, 37). While apostasy and blasphemy (another capital offense, as Khomeini's fatwa against British author Salman Rushdie reminded the world) may not routinely result in execution, it is noteworthy that reservations against the principle are almost never heard from clerics or commoners, and opposition to condemnations is rare. In Iran in 2002, when a conservative court sentenced a university professor to death for publishing a proposal for an Islamic "enlightenment," thousands of university students did take to the streets in protest. That such a sentence could be handed down at all is indicative of the gulf between the dogmatic and the rational in this historic and civilized society. It is also indicative of the power and prevalence of the dogmatic and the paucity and scarcity of the rational. Thus to express revulsion, or even disapproval, of acts of terror perpetrated in the name of Islam entails risks no "moderate" Muslims can afford to take and few mullahs or imams would encourage.

Importantly, Muslim rage is not directed only at the West in general or America in particular. Terrible, death-dealing conflicts have marked the Islamic realm as they have the Christian world, conflicts born of sectarian issues as well as other, more worldly causes. The war between Iran and Iraq in the 1980s, which may have cost as many as a million lives, pitted Shi'ite Iran against a Sunni-ruled Iraq, but it was not primarily a sectarian conflict. The invasion of Kuwait by Iraq in 1990 was over egress and oil, not religion. The strife that began in western Sudan in 2003 and ravaged the Darfur Province on the border with Chad was over land, not faith. Acts of terrorism directed against rulers (such as the 1981 assassination of President Sadat of Egypt) and others reflect the volatility of Islamic society. "In the early 1990s Muslims were engaged in more intergroup violence than were non-Muslims, and two-thirds to three-quarters of intercivilizational wars were between Muslims and non-Muslims. Islam's borders are bloody, and so are its innards" (Huntington, 1996). Twenty years later, the convulsions of Lybia, Syria and other Muslim states in the wake of the "Arab Spring" seem to confirm these assertions.

Islam, as a belief system, is six centuries younger than Christianity. It is worth considering where Christianity was in the 1400s, the time when Joan of Arc was burned at the stake after a church court condemned her of heresy. Roman Catholic ecclesiastical abuses were leading to the Protestant Reformation, and Martin Luther would soon enter the stage. The Spanish Inquisition was authorized by Pope Sixtus IV to combat apostate Jews and Muslims and to pursue heretics; thousands were burned at the stake and countless more were tortured and dispossessed. Still to come was the terrible strife between Catholic and Protestant forces with incalculable civilian casualties and immeasurable cruelty; Catholic armies could have taken a page from the Quran when it came to dealing with unbelievers. In their Bible, Christians found encouragement in the chapters of Deuteronomy (and elsewhere) to engage in unspeakable barbarity. The Enlightenment still was 300 years away.

In the centuries that followed the Reformation, Christians learned to accommodate their sectarian differences, to live in reasonable harmony with each other, to constitutionally separate church and state, to guarantee citizens not only freedom of religion but also from religion. Except for Northern Ireland, where the conflict has involved more than religion alone, Catholics and Protestants are not killing each other over their religious preferences. Europe has been described as being in a "post-Christian" stage as churches are emptying and congregations are ageing, but this trend is a matter of free choice and there is no coercion to reverse it. Christian fundamentalism has expanded on the religious and political maps of the United States, but it has no violent dimension. Some observers, including courageous Muslim intellectuals, suggest that Islam needs and awaits a Reformation of its own, and the Muslim world an Enlightenment that will obviate the frustrations and anger now pervading the culture. But Islam is not organized the way hierarchical Christianity is, with popes and bishops whose decrees can control, or at least influence, the most remote congregations. While there are ayatol-lahs and imams in the more structured Shi'ite sect, the overwhelming majority of Muslims are Sunnis, and Sunni clerics are masters of their mosques. No edict or decree will change this; it has taken the Saudi po-litical regime, not the religious establishment, to reign in some (by no means all) of the most extreme voices from the pulpits.

Certainly the successful al-Qaeda attacks of 9/11 were celebrated from some of these pulpits, but it was obvious to all that the world would change in unforeseeable ways in their aftermath, and that the overwhelming majority of people affected by these changes had nothing to do with what happened on that fateful day. President G. W. Bush pro-claimed from the rubble of the Twin Towers that "we" would find and punish the perpetrators, and since it was already clear that al-Qaeda's headquarters were in Afghanistan, it was obvious that a military cam-paign in that country was in immediate prospect. What was not obvious, at the time, was that another war was already under discussion in Wash-ington, or that, a decade and thousands of casualties later, neither cam-paign would have achieved closure.

AFGHANISTAN IN THE CROSSHAIRS

When Usama bin Laden returned to Afghanistan in 1996, the long-term investments made by Pakistan and by Saudi interests in the madrassas were bearing fruit. An Afghan *jihad* was in progress, driven by *taliban* (often translated as "students," although "seekers" is closer to the mean-ing of this name) trained in these schools in Pakistan and led by teachers who had turned them into religious fanatics. Also on the scene were an assortment of terrorists of note, including a physician named Ayman al-Zawahiri, who had been involved in the murder of Egyptian president Anwar Sadat and who was to become second in command of al-Qaeda to bin Laden. Another figure of note was Mullah Omar, the Taliban's so-called Commander of the Faithful, whose friendship with bin Laden helped make Afghanistan the terrorist redoubt it was.

When the Taliban (the capitalization reflects their role in the holy war) entered Afghanistan from their support bases in sympathetic Paki-stan, Afghanistan in the aftermath of the Soviet defeat was a country fractured into several dozen fiefdoms ruled by mujahideen warlords whose forces were in many cases well armed and who controlled all trade and transportation, levying tolls and tribute and taking the country back to feudal times. A skeletal government in Kabul had little power, even in the city's immediate hinterland, and essentially none in the more remote provinces. But it was clear that the areas where the largest ethnic group—the Pushtuns—were based had some semblance, at least, of stability. Thus when the mainly Pushtun Taliban seized control of the southern city of Kandahar and moved toward Kabul, they had considerable support in a population exhausted by decades of conflict. But even as their campaign to overpower the last resisting warlords continued, the Taliban began to impose upon their subjects the rules they had learned in their schools: the rules of sharia law. Women were compelled to wear burqas and prohibited from employment. Many of those working women "who had lost their husbands, fathers, and brothers in the war were forced to beg in the streets surrounded by their starving children" (Kepel, 2002). Entertainment in the form of television, radio, or live music was forbidden; on Friday the soccer stadium in Kabul was used not for games but for the public whipping of accused drinkers, the amputation of the hands of thieves, and the execution of condemned murderers. From his residence in Kandahar, Mullah Omar endorsed such practices and directed Taliban strategy against the remaining resistance. Surrounding him was the growing circle of Islamic terrorists who had converged on the safe house Afghanistan had become. Chief among them was Usama bin Laden, who would soon begin plotting the 2001 attacks on New York and Washington, D.C.

The physiography of Afghanistan and the country's relative location made it a receptive locale for such purposes. Nearly the size of Texas, compact in shape except the protrusion of land called the Vakhan Corridor extending eastward to touch China, mountainous in the center with countless valleys and steep-walled, strategic passes among which the Khyber Pass is the most crucial, and wedged between Shi'ite Iran to the west and Sunni Pakistan to the east, Afghanistan is the very definition of remoteness, isolation, and fragmentation (Fig. 6-3). In the east, rugged, forested terrain marks the border area with Pakistan, and in this cave-riddled topography Usama bin Laden was able to escape his pursuers, making his way (in all probability) across the border from Tora Bora and hiding on the rugged, wild, culturally closed Pakistani side. Here, in what maps call the Tribal Areas (Waziristan), bin Laden would have found allies to hide him and fighters to protect him. While he was a fugitive, bin Laden continued to issue taped exhortations and proclamations, sometimes standing out in the open, machine gun in hand. His

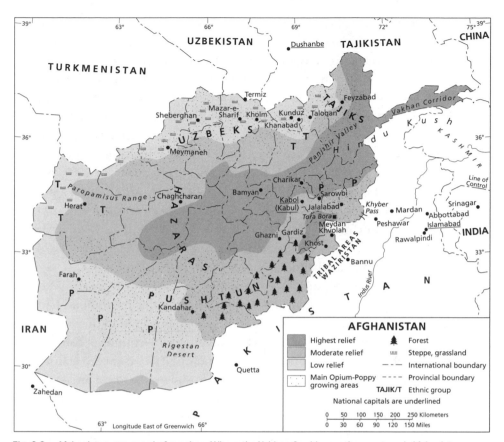

Fig. 6-3. Afghanistan, graveyard of empires. Where the Vakhan Corridor reaches eastward, Afghanistan touches China.

Allied pursuers asked American geologists and physical geographers to identify the terrain and even particular rock strata visible behind him in an effort to detect where he might be, but in Afghanistan or the Tribal Areas this was searching for a pin in a haystack. As is now known, bin Laden and members of his family and entourage managed not only to enter Pakistan but to hole up in a villa not far from Pakistan's capital Islamabad. Eventually American intelligence located him there and he was killed in an assault by U.S. Marines in 2011. Upon his death, Ayman al-Zawahiri became al-Qaeda's leader, but by then the terrorist organization had morphed into a far more diffuse and disjointed operation than it was a decade earlier.

In the war launched against the Taliban in September 2001, Afghanistan's dominant geography factor was the same one that had bedeviled the Soviets two decades earlier: the country's high relief and rugged to-pography, isolated refuges and remote frontiers. Taliban fighters and their allies were familiar with cave-riddled terrain in which modern weapons technology provided insufficient utility. Hundreds of Allied soldiers lost their lives in isolated valleys where military outposts proved vulnerable and where helicopters were at risk. Even the main "high-way" between the national capital Kabul and the "southern capital" Kandahar was a potholed stretch of gravel along which robbers lay in wait. If we had a map of traffic flows in the Afghanistan of 2001, we would find that more movement took place between Afghanistan's pe-riphery and neighboring countries, for example from western Herat to Iran and from northern Mazar-e-Sharif to Uzbekistan, than within Af-ghanistan itself (the heavily used and historically famous Khyber Pass between Kabul-Jalalabad and Pakistan's Peshawar confirms this).

Nevertheless, the campaign to oust the Taliban and to destroy al-Qaeda's refuge had initial success. Afghanistan under the Taliban had descended into medieval misery and the Taliban engaged in despicable acts of oppression and retribution while ravaging the country's cultural landscape (the dynamiting of two ancient and huge stone-carved statues of the Buddha at Bamyan raised a worldwide storm of protest and even a few expressions of "unease" from Islamic sources). Especially in the urban areas, the campaign against the Taliban had support; by mid-December 2001 the last Taliban urban stronghold, Kandahar, had fallen and a transitional government led by Hamid Karzai was installed. In January 2002 the military campaign known as Operation Anaconda drove what seemed to be the remaining Taliban forces and al-Qaeda groups into Pakistan. Schools reopened, UN-supervised elections took place, the transitional government led by Karzai was legitimized, relations be-tween the U.S. military leadership and locals were good, an Afghan-born U.S. Ambassador, Zalmay Khalilzad, was installed in Kabul, and when democratic elections in October 2004 resulted in Hamid Karzai's victory and his formal installation as president of the Islamic Republic of Afghanistan, it seemed that the only major unattained objective was the capture of Usama bin Laden. All else appeared to be within reach.

IRAQ REPRISE

While American interest was focused on Afghanistan and concern cen-tered on terrorist threats and national (and personal) security, something else was being discussed in Washington. Even before 9/11 it was no secret that President George Bush saw Iraqi dictator Saddam Hussein as a villain, not only because of Iraq's attempted annexation of Kuwait a decade earlier, but also because of his alleged role in a failed assassination plot on former president George H. W. Bush, architect of the multinational campaign that ousted Iraq from Kuwait but left Saddam in power. In the immediate aftermath of 9/11, he ordered his chief counterterrorism specialist, Richard Clarke, to search for a link between Saddam and the attacks on New York and Washington. "Over the months that followed, Bush and his aides pursued a policy that reflected their obsession. Instead of using their great power to crush a terror group responsible for devastating attacks on the United States, they turned against a dictator who, though odious and brutal, had never attacked Americans or threatened to do so (Kinzer, 2006).

In January 2002, in his State of the Union address, President Bush identified Iraq, neighboring Iran and remote North Korea as an "axis of evil" with which the United States would have to deal. Still, with the anti-Taliban campaign in Afghanistan in full swing and UN discussions regarding Iraq in progress, the prospect of a military attack on any of the three "axis" members seemed remote.

But in fact preparations for an invasion of Iraq were already underway. In his book *Intelligence Matters*, former Florida senator Bob Graham describes a briefing in Afghanistan by General Tommy Franks, following which the general, speaking privately to the senator, stated that "military and intelligence personnel are being redeployed to prepare for action in Iraq . . . what we are doing [in Afghanistan] is a manhunt" (Graham and Nussbaum, 2004).

It is possible to discern seven principal reasons for this diversion from Afghanistan to Iraq. Intelligence reports indicated that Saddam Hussein possessed chemical and biological weapons of mass destruction and might achieve nuclear capability as well. Iraq was seen as an immediate threat to Israel, which had strong allies in the Bush administration. Iraq was cooperating with terrorists in various ways, giving large awards to families of Palestinian suicide bombers. Iraq was defying UN resolutions and was doing business in contravention of UN sanctions. Saddam Hussein and his clique were perpetrating dreadful human-rights abuses on the general population. Iraq's oil industry was in a shambles and could once again become part of the global supply. And, as noted, George W. Bush had a personal issue with Saddam Hussein because of the dictator's foiled plot to assassinate his father during a visit to the region after the 1991 Gulf War.

Following difficult and divisive UN negotiations that erased much of the goodwill and support the United States had seen after the 9/11 attacks, American forces supported by British and Australian troops invaded Iraq from the south in March 2003. Turkey had rejected United States requests to use its territory to enter from the north. It was actually believed by some high-ranking members of the American administration that the invading force of infidel soldiers would be welcomed demonstratively by roadside crowds.

The sequence of events following the initial thrust is all too familiar. Meeting relatively little resistance, American forces reached the Baghdad area in a matter of weeks, entered the city and occupied Saddam's palaces, tore down his numerous statues, disbanded his army and police, witnessed a wave of looting and score settling in the general population, and began the effort to establish a post-Saddam administration. On an aircraft carrier off the coast of California, President Bush declared "mission accomplished."

But the real mission lay ahead, and for this the planners had poorly prepared—so inadequately that one wonders to what extent they were familiar with the historical or the cultural geography of Iraq. In many ways California-sized Iraq is the pivotal country in the Middle East, with 60 percent of the region's area and 40 percent of its population (demographic estimates range from 30 to 34 million). With the narrowest of outlets to the Persian Gulf, Iraq has six neighbors including Saudi Arabia, Turkey, and Iran, and significant historic and cultural ties with all of them. The country is heir to the earliest Mesopotamian states that arose in the Tigris-Euphrates Basin, and it is studded with matchless archaeological sites and museum collections, to which disastrous damage was done during and after the invasion from the combat and looting. With its major oil and gas reserves and large areas of irrigable farmland Iraq is also the best endowed of all the region's countries with natural resources (Fig. 6-4).

Iraq's physiography ranges from a mountainous east, where the rugged Zagros Mountains afford relatively few routes into neighboring Iran while the hilly northeast has less relief and greener countrysides, to a flat expanse of desert along the southern and western border with Kuwait, Saudi Arabia, Jordan, and Syria, where population is sparse and overland routes tenuous. Between these two extremes lie the coalescing basins

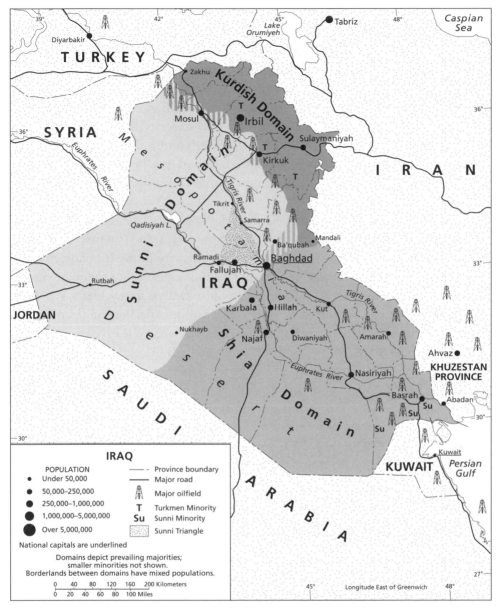

Fig. 6-4.   Iraq generalized. The notion of partitioning the country during the American intervention, perhaps temporarily, was mooted, then abandoned. It may yet come about.

of the country's two great lifelines, the Tigris and Euphrates Rivers with their distributaries and lakes, and here the population concentrates.

Iraq was one of those colonial creations with which European imperialists saddled the Muslim world, centered on historic Baghdad but encompassing peoples and cultures that have strong ties across its borders. About 25 million of Iraq's approximately 32 million citizens are Arabs (the others are mainly Kurds), but the Arab majority is divided on the basis of religion between Shi'ites in the south and Sunnis in the north. Shi'ites outnumber Sunnis by more than two to one, but it is the Sunnis who, through violence and intimidation, ruled Iraq throughout its modern existence.

Iraqi Shi'ites have had a difficult and complicated relationship with the Shi'ites of the Shia heartland, in neighboring Iran. The map shows part of this story: Arabs actually form the majority in the Iranian province of Khuzestan (capital Ahvaz), the partial annexation of which was one of Saddam's objectives during the war with Iran in the 1980s. What the map does not show is that Arab (Iraqi) Shi'ites and Persian (Iranian) Shi'ites do not see eye to eye. Iraq's Shi'ites adhere to a form of the faith known as Akhbari, which does not have strong political motivation and does not readily generate a political power structure. Iran's Shi'ites follow Usuli doctrines, in which the link between religion and politics is far stronger. As soon as the postinvasion disorder made it possible, Iranian Shi'ites crossed the border and began urging Usuli practices on the Iraqi faithful. One of those who heard that call was a young cleric named Muqtada al-Sadr, who assembled an armed force, holed up in a mosque in the holy city of Najaf, and contributed greatly to the chaos into which postinvasion Iraq descended.

But even Muqtada al-Sadr eventually agreed to participate in the political process, mindful of the order issued by the Shi'ites' grand ayatollah Ali al-Sistani, who proclaimed that failure to do so by any Iraqi is "a betrayal of the nation" that would be punishable by "burning in hell." Indeed, Akhbari doctrine goes far to explain why a comparatively small Sunni minority could mistreat the Shi'ite majority (which constitutes more than 60 percent of the country's entire population) for so long, even to the point of environmental terrorism when Saddam ordered the draining of the historic marshlands of the south, depriving tens of thousands of their traditional livelihoods. Terrible retribution was meted out to Shi'ites suspected of disloyalty following Iraq's defeat in the 1991 Gulf War. Mass graves found after the 2003 invasion contained the remains of hundreds of thousands, most of them Shi'ites. Nevertheless, Iraq's Shi'ites proved themselves repeatedly to be Iraqi Arabs first and Shi'ites second. During the Iran-Iraq War, there were no mass defections. None should have been expected in 2003.

The Sunni minority, too, has cross-border affinities, in its case to Syria. Saddam Hussein found refuge there after he was wounded in an assassination attempt on the Iraqi prime minister, before he participated in the coup that led to his dictatorship. The same political apparatus that keeps the minority Alawite sect in power in Sunni Syria, the Ba'ath ("Revival") Party, facilitated Sunni rule in Iraq. Ba'ath party structure is highly centralized and authoritarian, and discipline is rigid (Saddam was regularly reelected with 100 percent of the "vote"). Factional differences between the Syrian and Iraqi branches of the party precluded what leaders on both sides of the border wanted: eventual political union. In any case, the Euphrates River enters Iraq from Syria, and a corridor of Sunni habitat connects the two neighbors. That corridor, and the desert tracks beyond, proved difficult for American forces to police.

Figure 6-4 shows the area of Iraq populated primarily by Sunnis; note that the wider zone, along the Euphrates, becomes very irregular toward the northeast. In the hilly and mountainous north live most of Iraq's 6 million Kurds, a fraction of a landlocked, fragmented nation of perhaps 32 million partitioned by the boundaries of Iraq, Syria, Turkey, and Iran. Kurds are not Arabs and their languages are not related to Arabic, and wherever they live, they have sought greater autonomy from their rulers, whether Iraqis, Turks, Syrians, or Iranians. In Iraq this led to periodic retribution, and it is Saddam's use of chemical weapons against Kurdish villagers that left indelible impressions on the outside world. Following the 1991 Gulf War, the Kurds were given protection against Baghdad, and in their relative security they prospered as no other ethnic group in the country. Along the cultural border shown on Figure 6-4, however, there has been friction over land and rights between Kurds and Sunnis, and in addition this border area is home to smaller cultural communities such as the Turkmen and Assyrians, with whom relations are also tenuous.

Baghdad straddles the Tigris River, a vast urban agglomeration in some ways a microcosm of Iraq, with Sunni, Shi'ite, and Kurdish neighborhoods ranging from the well-off to the poverty stricken. To the east of the meandering waterway lies the slum once known as Saddam (now Sadr) City, populated by an estimated 3 million Shi'ites, where people were punished savagely for even minor infractions. To the west was the base of the privileged Sunnis, with key public buildings, Saddam's numerous monumental palaces, ornate mosques, statues, and other edifices of the regime. The American invasion severely damaged the city, which was for months afterward—and still, though less so, today—suffered from an inadequate and unreliable water supply, power outages, infrastructure damage that severely impeded circulation, inadequate medical facilities, and malfunctioning (or closed) schools. American planners established a secure "Green Zone" on the right bank of the river, where the U.S. Embassy and administration buildings were concentrated under strict (but not totally effective) security.

During the early period following the American intervention, called Operation Iraqi Freedom, the situation deteriorated drastically, at great cost to the civilian population and considerable loss to the military. A combination of terrorism (al-Qaeda saw and seized an opportunity it had not had earlier), Sunni-propelled insurgency, and Shi'ite resistance eroded security and put social distance between the occupiers and the occupied, worsening relations and eroding trust. A prison scandal involving mistreatment of inmates by United States soldiers did further damage. Bomb attacks on recruits hoping to join Iraq's new police and army, sectarian violence in the form of gruesome killings of pilgrims and worshipers, and resentment against the "de-Baathification program" (which was designed to stop Saddam's onetime allies from taking positions in the new Iraq but galvanized Sunni resistance) created a state headed for failure. Saddam Hussein's capture, imprisonment, trial, and execution only deepened intra-Iraqi animosities at a time when the country should have been heading toward representative government.

As the situation in "mission-accomplished" Iraq worsened, fateful decisions were made in Washington. Military leaders who had achieved much in Afghanistan were redeployed to Iraq. Ambassador Khalilzad was reassigned from Kabul to Baghdad. At a critical time, Afghanistan was relegated to the back burner. Despite a growing military force, violence in Iraq escalated to the point that, in 2006, the country was on the brink of civil war with tens of thousands of deaths on all sides in the struggle. Al-Sadr's Shi'ite militias played a major role in this, as did Sunni militants, aided by al-Qaeda operatives, former Baathists, and numerous splinter groups. In November 2006 the Iraq Ministry of Health published a report that estimated the death toll since the 2003 invasion at 150,000. Meanwhile, hundreds of thousands of Iraqis of all stripes including doctors, engineers, businesspeople, lawyers, artists and other professionals and their families fled when and where they could, becoming refugees in Jordan, Syria and elsewhere. In 2007 the United States government authorized a so-called "surge," adding some 30,000 troops to the occupation force of over 130,000 in an effort to stabilize the situation; in a parallel campaign, tribal leaders who had opposed (and fought) the Americans were now enlisted in the fight against an increasingly aggressive al-Qaeda.

When the surge brought some respite, negotiators redoubled their efforts to promote a coalition government in Baghdad, and in 2008 the Iraqi Parliament called for the U.S. to withdraw its troops from all cities and towns by the end of 2009 and for all U.S. forces to be out of Iraq by December 31, 2011. That commitment was echoed by then-presidential candidate Barack Obama, who ensured that this deadline was met.

The stated objective of the American-led invasion (apart from seeking evidence of a nuclear-arms initiative) was to end the cruel dictatorship of Saddam Hussein, to rid the country of his political heirs (including his two rapacious sons), to develop a system of representative government, to hold elections, and to leave behind a reconstructed, democratic, economically progressing Iraq endowed with rich energy resources that would stand at the heart of the Muslim world as an example of the advantages of freedom and self-determination. That this Trojan Horse of democracy might be seen as something quite different by regional states near and far may have occurred to President Bush and, more importantly, Vice President Cheney, but the evidence (from their own pens) is slight. Majority rule in Iraq would extend a Shi'ite axis from Iran toward Syria and Lebanon; such a government might promote Shi'ite causes throughout the realm. True democracy in Iraq would unsettle entrenched autocracies in this volatile part of the world. Iraq's economic opportunities could draw in an unfamiliar array of investors from China to Russia to Venezuela. Kurdish autonomy is an unwelcome development in government and military circles in Turkey. Iraq's dominant geography factor is its pivotal situation in its geographic realm, where the cost of success—if it holds—has been disproportionately enormous while the still-real risk of failure is incalculable.

FROM SELF-DEFENSE TO NATION-BUILDING?

While what was in effect a continuing war in Iraq took center stage in public debate in the United States, the campaign in Afghanistan had a

AFGHANISTAN AND VIETNAM:
ON PRESIDENTS AND PITFALLS

Hamid Karzai's victory in Afghanistan's disputed 2009 presidential election created a diplomatic and strategic dilemma that produced some troubling commentary by American officials and much strident criticism in the media. The then-Chairman of the Senate Foreign Relations Committee, John Kerry, in an interview from Kabul on *Face the Nation* on October 19, stated that the U.S. was facing strategic decisions "without an adequate government in place." Vice President Joe Biden was unsparing in his disparagement of Karzai, whose government and family are linked to corruption and drug dealing. In an October 14 column in the *New York Times*, Thomas Friedman lamented the "tainted government" of Afghanistan and the "massive fraud" engaged in by President Karzai to secure his re-election, arguing for a runoff to secure a more "acceptable" government to replace the one then in power, so as "to stabilize Afghanistan without tipping America into a Vietnam."

Comparisons between Afghanistan and Vietnam are frequently drawn these days, but the two contingencies are starkly different. Yet what happened in Vietnam in 1963 suggests caution in Afghanistan today. At that time, South Vietnam was in turmoil as the Viet Cong were gaining in remote northern rural areas; 12,000 American "advisors" were supposedly training South Vietnamese forces to shore up the South's defenses. South Vietnam's President Ngo Dinh Diem, facing growing Buddhist resistance marked gruesomely by public self-immolations by numerous monks, was unpopular with American policymakers. His autocratic methods, reputation for corruption, and harsh response to his religious opponents elicited severe criticism from American leaders and pundits. When

much lower profile. Expectations were also lower: Usama bin Laden's success in evading detection and capture suggested how difficult it would be to secure this Texas-sized country and to contain a Taliban resurgence given the refuges available to the ousted Taliban in neighboring Pakistan. Meanwhile, al-Qaeda seemed still to be capable of launching deadly attacks whose planners might or might not be operating from bases in Waziristan or elsewhere in Pakistan. In March 2004 terrorists struck the Madrid train station, killing nearly 200; in July 2005 three subway trains and a double-decker bus in London were bombed, causing 56 deaths. In August 2006 an al-Qaeda plot to simultaneously bomb several Transatlantic airliners was exposed before it could be carried out.

President Diem asked the United States government to reduce the number of American advisors in his country, he lost what little support he retained in Washington—and found his political base weakened at home.

On November 1, 1963 a military coup carried out by soldiers, some of whom had benefited from the presence of American advisors, overthrew President Diem, who was summarily executed. In official and media commentary in the United States afterward, Diem got little obituary solace. In South Vietnam, a so-called revolutionary council took power and inaugurated a fateful period of more compliant association with American policymakers.

American insistence on an electoral runoff in Afghanistan and Washington's apparent belief that President Karzai's opponent, if victorious, would have formed a less corrupt government may have been misplaced. The rules of political, social, and economic engagement in Afghanistan that have prevailed for centuries would not have been changed by an electoral runoff that might not only fail to alter the outcome but could risk chaos arising from the rekindling of hopes dashed and buried by Karzai's victory. Afghanistan remains a deeply divided country in which warlords, tribal chiefs, insurgents, brazen criminals, and a small cadre of courageous Kabul-based progressives are just some of the parties looking for their piece of the action; not for nothing do international monitors rank this as one of the world's most corrupt societies. Karzai, with his merits as well as faults, came to symbolize and stabilize the state; foreigners forcing a runoff risked leaving him either victorious but severely weakened or defeated with no guarantee of a superior successor. Add to this the alternate prospect of an adversarial "power-sharing" government and an ongoing political crisis, and it appears that one lesson of Vietnam, at least, went unheeded.

Nevertheless, Iraq continued to preoccupy the Bush Administration even as Afghanistan's numerous problems—increasing poppy cultivation and a growing narcotics trade, endemic corruption, a re-emergence of terrorist activity and even insurgency, ineffective central government and more—threatened the entire initiative launched after 9/11. Iraq's notorious—and oversimplified—three-way ethnic division had given rise to a public debate over the possibility that the country's temporary partition might offer a way toward eventual reconstitution, but a newspaper's random poll of citizens indicated that only one in seven could identify Iraq on a blank map of the region, suggesting that public opinion on this was, to put it mildly, not well-informed (de Blij, 2003). If

Iraq's geography was vague on Americans' mental maps, Afghanistan's was all the more so, even when the Iraq campaign began winding down and Afghanistan's problems and costs started to trouble policymakers. In truth, America's leaders did not do much better on Afghanistan than they had, as Secretary McNamara lamented, informed the public on Vietnam. On November 29, 2009, for example, Senator Carl Levin, Chair of the Senate Armed Services Committee, appeared on the CBS program *Face the Nation*, and host Bob Schieffer asked him what his impression was: "You're just back from Kabul, Afghanistan, Senator. What's your impression of the situation there?"

Answered Senator Levin: "Well, you don't have the kind of ethnic division there like you have in Iraq."

Mr. Schieffer did not follow up, but one wonders what viewers would conclude from this observation. Afghanistan's complex ethnic (and ethnolinguistic) mosaic, superimposed on its varied physiography, is its crucial geography factor. Nothing would be the way it is in Afghanistan without what David Isby calls the Vortex: the ethnic Pushtun domain that lies astride the country's border with Pakistan, creating a cultural continuity from Pakistan through Waziristan across Afghanistan to the border of Iran (Fig 6-5). Representing more than 40 percent of Afghanistan's 34 million people and some 14 percent of Pakistan's 180 million, they constitute an ethnic juggernaut of 60 million contiguous in the south but scattered and insular in the north. These are the insurgents. They grow the poppies. These are the Taliban. They are also dominant in Afghanistan's central government; their language, Pashto, is one of the country's two official tongues. And although Afghanistan's official name calls it an Islamic Republic, it is the Sunni Islam of the Pushtuns that dominates to the tune of 82 percent; Shia, the faith of the Hazaras and several smaller groups, is embraced by about 17 percent.

Indeed, to find Iraq-like ethnic-cultural discord in Afghanistan you do not have to go far from the center. The Mongol-descended Hazaras were brought under Pushtun rule by force in the late nineteenth century, but managed to fend off their powerful neighbors and get their Dari (Farsi) tongue recognized as the country's other official language. They had a brief heyday during the Soviet invasion and occupation, and their customs and traditions differ strongly from those of—let us call it what it is—Pushtunistan. But when the Taliban renewed Pushtun dominance and dragged Afghanistan back into medieval fundamentalism, the Hazaras were the country's biggest losers. Figure 6-5 indicates why.

Fig. 6-5. Ethnic components in Afghanistan's population. Note the Pushtun domination in the east and south, with outliers in the north, giving rise to the geographic name "Pushtunistan" as a player in the competition among interests here.

As the map shows, Afghanistan is bounded to the north by a set of neighboring states with ethnic names—and their ethnicity spills over the northern border. The second-ranking ethnolinguistic group is Tajik (about 27 percent), people with historic ties to Tajikistan. Numerically the Hazaras are next, and the Uzbeks follow. Although Uzbekistan's border with Afghanistan is short, the road crossing it leads directly to the important northern city of Mazar-e-Sharif, and the map reflects the considerable territorial extent of this Uzbek minority. When Taliban conquest threatened, a Northern Alliance of non-Pushtun forces tried to stop their advance. That conflict was in progress when the United States intervened after 9/11.

The campaign in Afghanistan suffered severely following the intervention in Iraq, and early successes and advantages were lost. Even in 2007, the total U.S. and NATO force numbered fewer than 50,000 soldiers, and Afghanistan's trained national army remained quite small. Meanwhile the Taliban began to use al-Qaeda-style tactics that had considerable success in Iraq: suicide bombing, kidnapping, taped and televised torture, and other heinous tactics. Thus the Taliban were able to some degree to reimpose the terror they had sown while in command a decade earlier, with serious impact on public confidence. For example, in 2008 they demanded that all mobile phone companies shut down at night, and when that did not happen, they destroyed signal towers to force compliance. Geographically, the Taliban focused their rural campaign in remote areas, especially in the far south of the country, and their urban strategy in Kabul. In the process, relations between Afghanistan and the Allies on the one hand, and Pakistan on the other, deteriorated: Pakistan was accused of harboring terrorist operatives and of not doing enough to stop Taliban fighters from crossing the border.

By 2009, the combined U.S.-NATO (but largely American) forces numbered over 100,000, and President Barack Obama, who as a candidate had forecast troop reductions and promised a review of the Afghanistan campaign, appointed Ambassador Richard Holbrooke as special envoy to Afghanistan and Pakistan to review the situation and to chart a new course. Part of this initiative was the co-option of so-called "moderate" Taliban who might be willing to participate in the political process. The country's so-called High Peace Council, headed by former president Burhanuddin Rabbani, received emissaries from the Taliban to discuss potential areas of common interest. Two years later, this was to cost Rabbani his life. On Tuesday September 20, 2011 he received a Taliban visitor who detonated a bomb contained in his turban. President Rabbani, a Tajik ally of Pushtun Afghan President Hamid Karzai, had been leader of the anti-Taliban Northern Alliance.

Predictably, President Obama's plans were subject to vigorous debate: following his agreement to add significantly to the troop total, there were some 150,000 Allied soldiers in Afghanistan in 2010, but insurgent attacks and terrorist assaults continued to increase. As the situation on the ground deteriorated, the president's commitment to begin troop withdrawal in 2014 seemed impractical and came under heavy criticism—even as Obama's supporters objected to his expansion of the force. The debate centered on a stark choice: should the United States seek some form of nominal control over the country, confront the Taliban and pursue al-Qaeda, assist in the consolidation of a representative government, train a national army and maintain a long-range presence, or should the United States abandon the campaign of nation-building, withdraw its armed forces, leave to Afghans the fate of elected government, accept the prospect of a Taliban resurgence in all its social implications and assist its Afghan allies by means other than a military presence.

Supporters of each proposition were able to marshal powerful arguments. On the "stay" side, these included (1) A military presence is essential because neighboring Pakistan has a substantial nuclear-weapons arsenal and is a weak state; failure of the Pakistani state would require American military intervention from its Afghan bases, (2) Not confronting the Taliban would risk reinstatement of the refuge created for al-Qaeda by the Taliban regime during its previous rule over most of Afghanistan, (3) The Taliban's revival in Afghanistan would spell doom for millions of women and girls in the repressive society that is its hallmark, and (4) the American presence in Afghanistan ensures a modicum of stability in this fractious region, improves local government and security, fights corruption and could replace opium production with other cash crops. On the "leave" side, the counterpoints included (1) The United States spent more than $1 trillion in Iraq and has budget woes that make the hundreds of billions of dollars required for this military presence and nation-building unaffordable, (2) India has more at stake than America when it comes to Pakistan's nukes, and an extremist Islamic takeover of these weapons would result in pre-emptive action by India whether the United States has a military presence in Afghanistan or not, (3) The pursuit of al-Qaeda should have been all-out and unrelenting immediately after 9/11, but now the al-Qaeda presence in Afghanistan is small and its putative Waziristan base is overshadowed by new and autonomous operations in Arabia and Africa, and (4) No external empire or force has brought enduring stability to Afghanistan, nor will the United States; the tragedy of women's fate here is by no means unique in the world.

Afghanistan's internal weaknesses may not be surmountable, but its attractions to foreign powers will remain undiminished. Take one final look at Figure 6-5: China has a short but crucial border with Afghanistan, and China's growing competition with India will make the Vakhan Corridor a crucial element in the political and economic geography of Asia. Already, China is exploiting one of the world's largest copper deposits at a mine not far from Kabul. India has interests in Afghanistan not only because of China's potential role but also because of Pakistan's dominant influence there. Pakistan in turn fears encirclement should India take a stronger role in non-Pushtun Afghanistan, notably through economic and other ties with the Tajik minority.

The geography factor is omnipresent in this complicated spatial puzzle. If the latest foreign mission in Afghanistan is never accomplished, it would not be the first time.

7

INTERPRETING TERRORISM'S
GEOGRAPHIC MANIFESTATIONS

AMONG THE MORE COMMON OBSERVATIONS MADE FOLLOWING THE successful terrorist attacks of 9/11 were predictions that the world would change forever in as-yet unforeseeable ways. On a crystal-clear September day shortly after the beginning of the new century, the skyline of one of the world's largest cities had been permanently scarred, the emblem of might in a superpower's capital had suffered a devastating assault, four loaded aircraft had been hijacked and used by suicide pilots as murder weapons, and more than three thousand civilians had perished. As buildings continued to burn and collapse in Manhattan and Arlington, the implications of the attacks were stupefying. Should follow-up attacks be expected? How could so complex a terrorist operation strike without warning? Where else was the United States so vulnerable? What security programs should be mobilized?

Certainly the world would change, but not in the ways many experts anticipated. More than ten years later, no remotely comparable terrorist operation had occurred. Al-Qaeda, the Afghanistan-based terrorist organization that had planned and funded the 9/11 attacks, found itself hampered by the Allied anti-Taliban campaign, its leaders on the run and its personnel dispersed into poorly coordinated cells in the Afghanistan-Pakistan border areas. Elsewhere, al-Qaeda squads operating autonomously mounted terrorist attacks that, while claiming hundreds of lives (as in the Madrid train-station attack of 2004 and the London subway-and-bus bombings of 2005) reflected none of the planning, dimensions, direction, funding, or coordinated execution that marked the 9/11 events.

Indeed, such "Islamic terrorist" attacks on non-Islamic targets were far outnumbered, every year, by intra-Islamic sectarian and political assaults. While other, loosely affiliated organizations adopted the al-Qaeda brand name (as in al-Qaeda on the Arabian Peninsula, al-Qaeda in West Africa, etc.), there was little evidence of a globally coordinated campaign despite the taped proclamations and exhortations emerging from bin Laden's hideouts prior to his elimination in mid-2011, and by his successor al-Zawahiri subsequently.

Compared to the hundreds of thousands of people killed by natural disasters in the ten years since 9/11, the toll from terrorism, dreadful as it is, remains low. The terror of terrorism has more to do with perception than reality: its randomness is frightening, its suddenness is fearful, its threat pervasive. Its psychological effect is far greater than its logistical impact. On Figure 4-2, we can gauge the riskiest places to live when it comes to earthquakes and volcanic eruptions. On any topographic map, it is clear where the hazards from coastal inundation by storms or tsunamis are greatest. But no map can forewarn against terrorist attacks. On April 19, 1995 a truck bomb destroyed the Murrah Federal Building in Oklahoma City, causing 168 deaths and more than 500 injuries. On July 22, 2011, a lone terrorist armed with explosives and ammunition bombed government offices in Norway's capital of Oslo and shot nearly 70 people attending a meeting on a small island nearby. Neither Oklahoma City nor Oslo had ever appeared especially vulnerable on anyone's mental map of terrorist risk.

But not all terrorism is so random; terrorist actions have tended to concentrate temporarily in locales where their tactics promised results. A map of terrorist activity in the United States after the Civil War shows the Ku Klux Klan of defiant white Southerners using violence to intimidate supporters of Reconstruction. A map of the "British Isles" between the two World Wars would show a clustering of attacks launched by members of the Irish Republican Army (IRA) that put targets in Northern Ireland and the Irish Free State as well as Britain at risk. In 1939 alone, an attack in Coventry killed five even as others in London, Manchester, and Birmingham caused major property damage. That campaign, of course, continued after the Second World War and brought the United Kingdom its first wave of terrorist mayhem, accruing more than 3,000 casualties.

During the 50 years straddling midcentury, the great majority of terrorist operations were of this kind. Their names were known worldwide and their geographic identities tended to be strong and local: the Red Brigades of Italy, the Baader-Meinhof Gang of (then) West Germany, the FALN of Puerto Rico, the ETA of Spain's Basque Country, Peru's Shining Path, FARC of Colombia, the Tamil Tigers of Sri Lanka. If their tactics were similar, their goals were not, ranging from sheer anarchy to territorial conquest. Each terrorist organization had its heyday, but few achieved their stated goals (not all proclaimed a clear objective). Some came close: FARC temporarily succeeded in creating an "insurgent state" within Colombia, and Sri Lanka's Tamil Tigers controlled a part of northern Sri Lanka until a government offensive defeated them. But, as many observers of the phenomenon have noted, other terrorist groups were simply in it for the sake of intimidation and violence. Comparatively few, from ultra-left to ultra-right, proclaimed ideological objectives to justify their actions.

Yet before midcentury a campaign supported by terror did succeed: the effort by Jewish nationalists to drive the British occupiers out of Palestine and to establish a state envisaged to lie astride the Jordan River. Their leading terrorist organization, named Irgun for short, was highly disciplined and virulently anti-Arab. On July 22, 1946 the Irgun, then led by future Israeli Prime Minister Menachem Begin, blew up a wing of Jerusalem's King David Hotel, killing 91 soldiers and civilians including British, Jewish, and Arab citizens. Irgun had a reputation for ruthless terror: on April 9, 1947 its commandos attacked an Arab village and killed all of its 254 inhabitants. The United Nations voted later in 1947 to partition Palestine into a Jewish and an Arab state; British withdrawal began and Israel was established as a sovereign state on May 14, 1948. Israel's Arab neighbors not only rejected partition but attacked the new state, precipitating a conflict in which Israel gained substantial ground. It was a geographic realignment that would form a looming backdrop to a new era of terrorism.

TERROR ON THE MAP

If it is impractical to devise maps of future vulnerability to terrorism, it is nevertheless possible to record past events and to discern patterns of change. During the period of rapid decolonization in Africa and Asia in the decades following the end of World War II, insurgencies mobilized by groups designated by colonial regimes as terrorist organizations gave a new, global dimension to terrorism. Seen as freedom fighters by locals, groups such as Kenya's Mau Mau and Vietnam's Viet Minh gained the kind of notoriety formerly reserved for European domestic terror groups.

One difficulty in mapping terrorist activity (apart from designation) lies in the sheer number of incidents terrorists generate. Germany's Baader-Meinhof Gang, also known as the Red Army Faction, was formed in 1968 and engaged in bombings, arson, kidnappings, and assassinations. A content survey of German media during the ensuing decade suggests that this outfit engaged in more than 300 attacks, many of them high-profile incidents including airplane hijackings. But this number pales against that of Italy's Red Brigades, who engaged in about 14,000 terrorist attacks in just one decade after their founding in 1970 (Moss, 2008). Most of these attacks caused property damage rather than casualties, but several had the desired publicity effect, notably the gruesome murder in 1978 of former Italian Prime Minister Aldo Moro, whose body was found stuffed in the trunk of a parked car in Rome.

In the 1970s, therefore, maps of terrorist activity showed European countries with the highest incidence, higher still than the growing rate in the Middle East and the decolonizing world. Painstaking police work, especially in Germany, Italy, and France, was nevertheless having its effect, and in the 1980s even the most notorious Europe-based terror organizations were being dismantled. Meanwhile, however, signals of the future had already appeared. Groups like the Red Army Faction and the Red Brigades had notions of undermining national governments in favor of Marxist regimes propelled by a "revolutionary proletariat." But quite another kind of motive brought eight Palestinian terrorists to the Munich Olympic Games in West Germany to kill 11 Israeli athletes and a local police officer.

Until that fateful September day in 1972, German concerns over security and terrorist threats had focused on the Red Army Faction and its still-unclear ties to other terrorist organizations. As in Spain, Italy and elsewhere in Europe, terrorist groups, if they proclaimed their goals at all, had domestic territorial or ideological objectives, not global aspirations. A map of contemporary terrorist activities or incidents would show spikes in all European countries that outnumbered those anywhere else, including in and around the young state of Israel. In 1967, Israel had fought a second round of war with its Arab neighbors, producing another victory with major territorial gains in the Golan Heights (taken from Syria), the West Bank (Jordan) and the Sinai Peninsula and Gaza

(Egypt). What was left to the defeated Arabs was the weapon of the weak: terrorism. And while the number of terrorist incidents targeting Israel and Israelis rose markedly in the aftermath of the 1967 war, none would achieve the global media coverage the Munich assault did. If you were mapping the pattern in color, you would use the color for the Israeli-Arab conflict not only in the Middle East, but now, for the first time, in Europe as well.

Thus the name of the Palestine Liberation Organization (PLO) gained prominence not only as an umbrella political organization claiming to represent the several million Palestinian Arabs including those who resided in the Palestine Mandate before the creation of Israel as well as their descendants, but also, through its internal and associated factions, as a terrorist syndicate. Formed in 1964 to coordinate the various Palestinian resistance movements, its profile rose after the 1967 War; its key terrorist organizations were the Black September movement of Fatah (which was responsible for the Munich operation) and the Popular Front for the Liberation of Palestine (PFLP). The latter was to become notorious for its hijacking and destruction of Israeli commercial airliners in the late 1960s and 1970s.

Now the hotbed of terrorist activity lay in the Middle East, although the PFLP pursued Jewish targets wherever they might be vulnerable. On June 27, 1976 Arab terrorists hijacked an Air France aircraft after it left Athens; the plane landed in Entebbe, Uganda where the passengers were rescued by Israeli commandos in a dramatic operation a week later. On August 11 Arab gunmen fired on passengers about to board an El Al airliner in Istanbul, killing four. As the circle of terrorism widened, reports of attacks became routine, increased security measures at airports and other facilities affected travelers virtually everywhere, and the Arab-Israeli conflict had become a global issue.

But it was not primarily a religious issue. Fundamentally the conflict was over territory and boundaries, ancestral rights and future cohabitation. Halting progress toward accommodation continued to be punctuated by dramatic terrorist acts, among which the hijacking of a cruise ship carrying 400 passengers and crew by four Arab gunmen in 1985 seemed a portent of things to come. Nor was the terrorism all one-sided. Terror attacks evoke revenge, and individual Israelis as well as the Israeli state engaged in retaliation. Israeli attacks on Arab refugee camps in Beirut in 1982 and 1983 brought charges of state terrorism, but in truth

Israel's neighbor Lebanon was in civil war. This was a time of momentous change in the geographic dimensions of terror: a widening campaign was in prospect.

FROM JETS TO *JIHAD:* CROSSING THE ATLANTIC

Hijacking jetliners and taking hostages in the name of the PLO gained attention and notoriety, but did little to change the calculus on the ground. Lebanon's chaos on Israel's doorstep involved a new set of combatants entirely, propelled by new and potentially more powerful motives. Lebanon, once a stable multicultural country whose capital, Beirut, was the region's financial headquarters, fell victim to internal and external forces and collapsed into chaos during the 1970s. Israel intervened to protect its citizens and to pursue Palestinian opponents; Shi'ite and Sunni Arabs, Christians and Druze fought among themselves; Syria policed a cease-fire. Outsiders were there to protect their interests. The year 1983 would be fateful: more than 50 people were killed by a bomb at the United States Embassy in Beirut, and 241 U.S. Marines lost their lives in a suicide attack. In a separate assault, 58 French soldiers died. Chapter 6 chronicled the slow but growing awareness of something called Islamic Holy war (Islamic Jihad), still a vaguely defined organization whose name had begun showing up on intelligence reports but whose operatives' names and bases could yet be identified. One indication that things were changing came from that 1985 hijacking of an American airliner on a flight from Athens to Rome by two gunmen who were Shi'ite members of Islamic Jihad and who demanded that Israel release more than 700 mostly Shi'ite prisoners who had been captured in Lebanon. The PLO and its operatives were Sunni Muslims but their political and territorial goals, not their sectarian affiliation, drove their actions. The breakdown of order in Lebanon now thrust sectarianism into the mix, notably the aggressive role of the Shi'ite minority that would become the core of the terrorist organization known as Hizbullah.

Not only are the Lebanese a fractured nation; they also have, and always had, worldwide linkages that proportionately exceed those of any other country in the region save Israel. A long history of emigration has taken Lebanese diaspora—Christian, Shi'ite, Sunni and others—to many parts of the world. In the colonial era no African country was without its Lebanese shopkeepers. An estimated 20 percent of all Argentinians have recent or distant Lebanese ancestry including a former President, Carlos Menem. The geography of Lebanese dispersal would become a key factor in the global reach of Hizbullah—aided and abetted by the political cornerstone of Shi'ism, the Islamic Republic of Iran.

Given these geographic pointers, it is all the more surprising that what seemed to be inevitable from the sequence of maps of terrorist activity from 1980 to 1990—its crossing the Atlantic Ocean to the Americas—was not anticipated as it should have been. By the early 1990s, Hizbullah had become a force in Islamic Jihad, appealing to Shi'ite minorities throughout the realm and beyond for support. When the first major Islamic-terrorist attack occurred in the New World, the 1992 bombing of the Israeli embassy in Buenos Aires, Argentina, suspicion fell almost immediately on Lebanon-based Hizbullah in collusion with staff members of the Iranian embassy there, but no concrete evidence could be assembled because of diplomatic immunity.

Should this successful attack have forewarned U.S. counterterrorist workers of the next terrorist assault in the Americas? No link appears to have existed, but in 1993 Americans received a warning in the matter of targets when a massive truck bomb exploded in the parking garage of the World Trade Center in New York, the obvious objective having been the toppling of one of the world's two tallest buildings. The blast killed six persons and injured more than 1,000. Investigators soon uncovered the existence of an Islamic-fundamentalist network in the United States led by an Egyptian cleric whose aim was to commit numerous other acts of urban terrorism in America.

And then, in 1994, Hizbullah struck once again in Argentina, this time in the heart of the capital, a devastating bombing of the Jewish Community Center on the city's fashionable Florida Avenue, resulting in 120 deaths and hundreds of wounded. Following this attack, pressure to investigate and solve both the 1992 and 1994 attacks heightened, but reportedly, high-level interference slowed the process down. In 1998, however, Argentinian judicial and intelligence officials got a break when an Iranian defector in formal testimony implicated senior members of the Tehran government including the president, the minister of foreign affairs, the head of intelligence, the son of the ayatollah Khomeini and the Iranian ambassador to Argentina at the time of the attacks. Additional information, gathered by Argentinian intelligence, showed Iranian officials entering and leaving Argentina under false names around the time of the bombings (Rohter, 2002a).

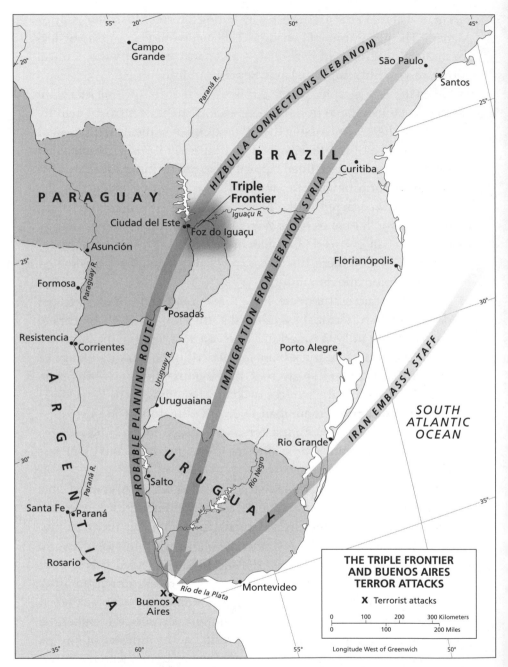

Fig. 7-1.   The Triple Frontier in southern South America, apparent staging area for the first major Islamic terrorist strike in the Western Hemisphere.

The ongoing investigation uncovered a geographic dimension that continues to matter today: the apparent role of the so-called Triple Frontier where Argentina, Brazil, and Paraguay converge (Fig. 7-1). This area, called the Triborder Area by the United States State Department, has an Arab population reported to exceed 25,000 and is described as "teeming with Islamic extremists and their sympathizers, [where] businesses have raised or laundered $50 million in recent years" (Rohter, 2002b). Records of telephone calls from the Iranian embassy to Hizbullah members revealed an organizational link between the Iranians and an apparent base in this remote area. Whatever its role, *New York Times* journalist Larry Rohter reported, the Triple Frontier had become a haven for militants, fugitives, forgers of passports and credit cards, telephone switching operators, money launderers, and enforcers. In the process, the cultural landscape was changing rapidly, with a proliferation of mosques and Islamic prayer houses and an infusion of Arab-language radio and television stations broadcasting round-the-clock sermons from Lebanon, Saudi Arabia and elsewhere in the Muslim world.

In the aftermath of the terrorist attacks in Buenos Aires, the Triple Frontier appeared to pose a threat as a possible staging area for a wider terrorist campaign, especially when a map of the area was found in an al-Qaeda safe house in Kabul, Afghanistan. But Hizbullah, having achieved its goals in South America, now turned to other objectives.

THE ISLAMIC FRONT IN AFRICA

Islam spread along the northern shores of Africa by caravan and along the eastern coast by boat, establishing itself in modern-day Egypt before the end of the seventh century and reaching Morocco before the end of the eighth. The faith's rapid diffusion transformed not only what we call the Middle East today but also North and East Africa, reaching Dar es Salaam, Tanzania by the eleventh century—about the same time it penetrated modern China after sweeping across Central Asia. From their bridgehead in North Africa, the Moors (or *Mors*, hence *Morocco*) invaded Iberia, where flourishing Islam left indelible imprints on the cultural landscape.

While ruling al-Andalus, the Muslims also looked southward, to West Africa across the Sahara. There, on the savannas between the forests of the coast and the desert of the interior, lay a tier of thriving African states in control of trade and traffic. From what is today Senegal in the west to northern Nigeria in the east, these states benefited from what economic geographers call "double complementarity": the peoples of the steppes to the north needed goods available from the forests of the south (starches, spices, animal products, building materials) while those of the forests wanted items from the interior such as leather and salt. These goods were traded on the bustling markets of the savanna states, and the Niger River ("where the camel met the canoe") was the Mississippi of West Africa. Through these markets, attracting the attention of the Muslims of Morocco, moved a considerable quantity of gold and gemstones.

Stable, durable, and in control of large populations and territories, the West African states such as Ghana, Mali, Songhai, and others were tempting targets for the powerful Muslims of the north. Caravans carried not only Moroccan leather but also Muslim proselytizers, and before long the kings and chiefs of West Africa's savanna states were converted to Islam and were ordering their subjects to follow suit. This led to a momentous cultural reorientation, because now the savanna corridor between forest and desert stretching across Africa from west to east became a conduit to Mecca. Ghana fell apart when the Muslim invasion was backed up by armed force, but Mali survived, and from there, gold-laden annual pilgrimages involving tens of thousands of the newly faithful walked eastward, making Khartoum a gathering point before heading for their final stage to Mecca. Khartoum became a West African outpost, its population swelled by pilgrims who never made it to Mecca and others who stayed there on the way back.

Christianity, with its six-century head start, did not make an impact in West Africa until colonial times, but in Africa's "Horn," where Ethiopia lies today, several African states had accepted Christian beliefs: Kush, Nubia, and Axum among them, the last giving rise to the Christian dynasty that eventually shaped modern Ethiopia. When Islam reached this area, it could not overpower these kingdoms, and Christianity survives there to this day, virtually encircled by Islamic societies.

Islam's southward march was halted by the arrival of European colonists, who established beachheads along the West and East African coast and moved inland, taking control not only of peoples who practiced traditional African religions but also those who had adopted Islam. The modern map of West Africa bears witness to this episode: boundaries between former British, French, and German possessions run from the coast into the interior, sometimes for dozens, elsewhere for hundreds of miles. Britain's Nigeria extended from the coastal forests to the interior desert, and its girdle of boundaries threw together Muslims and non-Muslims alike. Then the colonists set about promoting Christianity among their subjects, so that Nigeria today has a Muslim north and a mainly Christian south (traditional African religions do survive in Africa's "Christianized" areas), creating severe tensions that continue to lead to recurrent riots and political strains that could eventually endanger the state.

When the colonial powers gathered in Berlin in 1884 to finalize their partition of Africa, they paid little heed to the religious divide that stretched across the continent from Guinea in West Africa to Kenya in East Africa. Upon decolonization, not only Nigeria but other West African states, as well as Chad, Sudan, and hemmed-in Ethiopia to the east, found themselves with regional-cultural contrasts fraught with problems. In the process, the colonial powers created a regional problem whose consequences they could not foresee. The Islamic Front has become a zone of conflict that now threatens the cohesion of countries and facilitates the actions of terrorists and insurgents.

Undoubtedly the costliest impact of the divisive Islamic Front has been in Sudan, whose Muslim government waged a long, bitter war against the southern provinces where African Christian and animist communities are in the majority. The cost in terms of casualties and dislocations will never be known; estimates of the death toll over more than three decades of conflict are in the hundreds of thousands. The fundamentalist-dominated Sudan regime wanted to impose Islamic (Sharia) law in the South; Southerners wanted independence or, failing that, protection and guarantees against Khartoum's cultural domination. When the considerable dimensions of oil reserves in the contested area were recognized, a settlement seemed ever more remote, but the unlikely did happen: following a referendum Sudan's southern provinces seceded and, in 2011, became the sovereign state of South Sudan. As Figure 7-2 shows, the new African state's northern border closely approximates the location of the Islamic Front.

In West Africa, the Islamic Front's proximity to Côte d'Ivoire proved costly as well. Long a stable and relatively prosperous country, Ivory Coast had had a northern Muslim minority ever since its territory was bounded. But in postcolonial years, Muslim herders and farmers from neighboring Burkina Faso to the north have been crossing the border

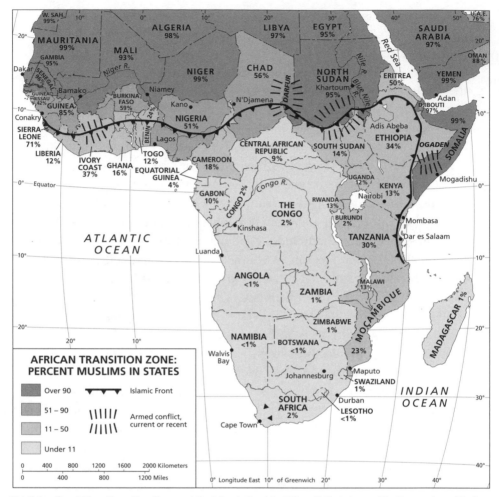

Fig. 7-2. The African Transition Zone and the Islamic Front in Africa. Strife between Muslims and non-Muslims occurs intermittently along this cultural "fault line"; al-Qaeda affiliates operate from the Somali sector to Nigeria and Niger.

and settling in rural areas of the country. Initially there were occasional skirmishes over crops damaged by Muslim pastoralists' cattle and land taken by immigrants, but then the growing strength of the Muslim minority began to translate into political power. When the country's long-term ruler died and a Muslim politician sought the president's office, he was disqualified by southerners and civil war broke out, ruining the cocoa-based economy and precipitating French intervention. The French force was accused of favoring the Muslim northerners, the large French expatriate community was set upon by mobs, and even the modern, once-thriving port of Abidjan was engulfed by violence. The political crisis was not resolved until 2011. The effects of the Islamic Front had reached the sea.

If the breakdown of order in Côte d'Ivoire was a shock to observers of the African scene, simultaneous events in Liberia were even more surprising. Liberia has long been in the grip of ethnic conflict, but in late 2004 reports from the capital, Monrovia, described serious clashes between Muslims and Christians in several parts of the country, the first instance of such religious strife in the country's history. As Figure 7-2 shows, Liberia lies to the south of the Islamic Front, but the Islamic component of its population has been rising and is now 12 percent. Still, Liberia's Muslim minority is scattered in the main towns, not in the northern region of the country. Whether the Muslim-Christian clashes are an isolated event or a signal of more serious problems is not yet clear.

But the largest and most tempting target in West Africa—and, surely, in Africa as a whole—is the realm's most populous state, Nigeria. Its population just about evenly (and regionally) divided between a dominantly Muslim north and a mainly non-Muslim south, its history replete with sectarian conflicts and its export economy dominated by oil, most of which is bought by the Great Satan across the Atlantic, Nigeria is a crucial prize for extremists bent on sowing terror and destabilizing governments.

Islamic issues have roiled Nigerian society ever since the Muslim advance halted in its "Middle Belt" and European colonialists encased Muslim as well as non-Muslim within a boundary that extended from ocean to desert. After independence, Nigerians negotiated a federal constitution establishing three dozen States, but weak, corrupt, and eventually dictatorial governments sowed seeds of dissatisfaction. Following the return of democracy in 1999, twelve northern States unhappy with southern domination decided to proclaim strict Islamic Sharia law, causing civil disorder (the ancient capital of Kaduna was devastated) and leading to the departure of thousands of Christians, intensifying the sectarian fault line shown on Figure 7-2 (see also Figure 11-3).

In 2009 an Islamic terrorist group called Boko Haram (the name means "Western Education is Sinful"), apparently affiliated with al-Qaeda, began a campaign that brought assassinations, car bombs, and other forms of violence to Nigeria. In 2010 its operatives attacked the central jail in Bauchi, releasing 700 inmates including 150 Boko Haram members imprisoned there for earlier offences; by mid-2012 Boko Haram had become a significant factor in Nigeria's security concerns.

THE VIOLENT HORN

During the first decade of the century the Horn of Africa and its neighboring areas became a cauldron of cultural conflict and terrorist activity.
Here the Islamic Front has significance beyond Africa itself: terrorist allies operate in Yemen on the Arabian Peninsula as well as Somalia. The
intervening waters of the Gulf of Aden have become the scene of the
specialized form of terrorism called piracy, making historic maritime
routes unsafe for commercial and tourist travel. In unstable Yemen, al-
Qaeda on the Arabian Peninsula has long taken advantage of the weakness of the state and the mountainous hideaways of the remote interior.
In failed-state Somalia, the terrorist organization known as al-Shabaab
has launched deadly attacks within and outside the country, including
the bombing of a crowd watching the televised World Cup football (soccer) final in Kampala, Uganda, killing nearly 80 fans. In taking responsibility for these murders, al-Shabaab issued a statement saying that the
assault was in retaliation for the peace-keeping forces Uganda had sent
to Somalia.

As Figure 7-2 shows, the Islamic Front loops into the Horn of Africa,
coinciding very roughly with the border between mainly Muslim Eritrea
and dominantly Christian Ethiopia before dividing the latter into a Muslim east (Ogaden) and a Christian west. The Muslim east is the historic
home of the Somali people living on the Ethiopian side of the border
with the failed state of Somalia that, as the map indicates, is virtually
100 percent Muslim. Ethiopia's Christianized core area, centered on the
capital of Adis Abeba, lies in the highlands, a natural fortress that has
protected the country in the past and from where the founding emperor,
Menelik, extended its power over the encircling plains. In the process
Ethiopia's Christian rulers gained control over the Ogaden with its
Muslim Somali population and its leading city, Harer. Thus the Somali
found themselves on both sides of a superimposed border, but after the
colonists withdrew it mattered little. Clan allegiances and divisions, not
invented boundaries, dominated life. Somali pastoralists drove their
herds across the borders in pursuit of seasonal rains as they had for centuries, and in Somalia they fought among themselves over influence and
dominance.

Long before Somalia became a hearth of Islamic terrorism, Kenya had
been experiencing the effects of terrorist activity based on the Muslim
side of the Islamic Front. Somali *shifta*, marauders on the move, had been

attacking targets in northeastern Kenya ranging from game lodges to archeological digs and from cattle owners to bus drivers for many years. They struck as far west as Lake Turkana, disappearing into the Ogaden on their way back to Somalia; they traded their sharp-edged *pangas* for guns, and they always were a menace. But they did not have an ideological agenda: that dimension only appeared during the 1990s, when East Africa became the stage for al-Qaeda's attacks on United States embassies in Nairobi and Dar es Salaam and subsequent assaults on a tourist hotel near coastal Mombasa and an Israeli airliner departing Mombasa Airport. As in the case of Bali, note the juxtaposition of refuge and target: Western interests in East Africa's cities lie short distances from areas as tribal as Pakistan's Waziristan.

A FRONT DIVIDED

The political geography of Africa's Horn and the neighboring Arabian Peninsula has become a matter of growing concern for several reasons. On the African side, the breakdown of the Somali state has created fertile ground for terrorist movements, and links between al-Shabaab and Somali-American militants on the one hand, and with al-Qaeda on the other, clouded the future. Al-Shabaab became increasingly radicalized as foreign extremists took operational and tactical control of the organization; one of these prominent foreign leaders was Omar Hammami, who made videos recruiting Somalis in America to join the terrorist campaign. On the other side of the Gulf of Aden, another American-born militant, Anwar al-Awlaki, had become the mouthpiece of al-Qaeda in the Arabian Peninsula (AQAP), calling for attacks on American targets. While the Yemeni regime of President Saleh opposed and pursued al-Qaeda, al-Awlaki managed to instruct and equip the Nigerian "underwear" terrorist and had a hand in other plots; he also counseled the Muslim U.S. Army officer who later killed 13 people at Fort Hood, Texas. In April 2010 the Obama Administration put al-Awlaki on its capture-or-kill list and he was killed by a strike from a drone in September 2011.

Yemen, of course, was the scene of one of al-Qaeda's most dramatic early successes, the October 12, 2000 bombing of the destroyer U.S.S. Cole when the ship was in the port of Aden for a refueling stop, leaving 17 Americans dead and nearly 40 injured. Bin Laden allegedly was the architect of this attack, and soon after 9/11 the United States sent troops to Yemen to track down members of al-Qaeda as part of the "War on Terror." But Yemen presented a difficult combination of conditions under which to pursue terrorists. In 2008, amid AQAP threats against the Saudi and Yemeni regimes and calls for the establishment of an Islamic caliphate, a suicide attack on the U.S. Embassy killed 17. Many observers regarded AQAP a stronger force and a greater threat than its predecessor in Afghanistan-Pakistan, and advocated a shift in priorities to this ever more dangerous area on the Arabian Peninsula.

But another conflict, potentially even more consequential because of its sectarian context, buffeted Yemen from a different direction. About 42 percent of Yemen's 25 million citizens are Shi'ites, and for years the Sunni regime fought al-Houthi rebels in their mountainous domain in the country's northwest, on the border with Saudi Arabia. A peace agreement was reached in 2010, just before the "Arab Spring" engulfed Yemen and disorders broke out anew—but by that time a significant development had occurred.

Across the Red Sea from the domain of the Houthi lies Eritrea (see Figure 7-2), with a Lebanon-like multicultural mix of Muslims (with a Sunni majority), Christians (including Coptic, Catholic, and Protestant), and smaller groups. Muslims slightly outnumber Christians in the population of less than 6 million. After a long struggle for independence and a promising start in 1993, Eritrea plunged into war with its neighbor Ethiopia and descended into autocracy. A 2007 UN report accused Muslim-dominated Eritrea of helping terrorists in Somalia; a Human Rights Watch Report dated 2009 described Eritrea as one of the most closed and repressive states in the world. Note that, despite the sizeable Christian minority, the Islamic Front in the Horn virtually coincides with the Eritrea-Ethiopia border.

Although details remain obscure, persistent reports of a Hizbullah presence in Eritrea drew attention when Houthi fighters in Yemen reported having been trained in Hizbullah-run camps in Eritrea. The treatment of Shi'ite minorities on the Arabian Peninsula is a priority matter for Iran's activists, second only to their enmity toward Israel. Iran's media made much of the Houthi rebellion in Yemen (in which the Yemenis got help from the Saudis), and the opportunity to support a Shi'ite cause from secure bases across the Red Sea may have proved irresistible. If so, the Islamic Front has more than one dimension in the turbulent Horn.

Is the Islamic Front static or mobile? The evidence from Côte d'Ivoire to Kenya suggests the latter. But even as the Islamic Front moves southward, African countries well beyond its leading edge form obvious opportunities for Islamic expansion. The horrific experiences of Rwandans, Congolese, and the citizens of other malfunctioning African states (or parts of states) may be so strongly associated with Christian contexts—many Rwandans were herded into churches and killed with the conniv-ance of ministers—that Islam appears as a hopeful alternative. Although data may be unreliable, it is noteworthy that Rwanda two decades ago reported that 2 percent of its population was Muslim; today it is report-ing 13 percent, a huge increase in a Roman Catholic country.

Figure 7-2 shows the percentages of Muslims in the populations of African countries, and the Islamic Front effect is obvious from these data. In general, the closer to the Islamic Front, the higher the Muslim com-ponent. Tanzania's high percentage, much higher than Kenya's, results in part from its merger with Zanzibar, since almost all Zanzibari are Muslims. But even so, Tanzania's coastal zone, including Dar es Salaam, is strongly Muslim, and may become a regional extension of the Islamic Front. Note that countries in approximately the same latitude—the Dem-ocratic Republic of Congo, Malawi, Zambia, Angola—report a minimal Muslim presence.

Which leaves the country farthest from the Islamic Front but promi-nent in Islamic history: South Africa. More than 350 years ago, the Dutch began to bring Muslim Indonesians to Cape Town as workers, and these "Malays," as they are often called, built the first prayer houses and es-tablished a base for the faith. Over time, these Malays became part of the larger population referred to as the Cape Coloured community, and Islam spread into it: this, during the decades of *apartheid*, was one way to reject the churches that preached oppression. Small, graceful mosques were part of the cultural landscape of Cape Town's historic District Six when it was demolished under apartheid laws to move its nonwhite residents farther from the city center.

Today South Africa's population is about 3 percent Muslim, and some of the country's 1 million Muslims are South Asians, descendants of a later immigration now living in Kwazulu-Natal, in the city of Durban and its hinterland. But Islamic activism, and some incidents of terrorism, center on Cape Town. Soon after South Africa's democratic transition, which unleashed freedoms unheard of under the repressive apartheid regime, an Islamic group began issuing fatwas against pornography, homosexuality, drug use, and other "excesses" of freedom; warnings against commercialism were followed by the bombing of a Planet Holly-wood in a new waterfront shopping center. Effective police work rolled up the group responsible, but a low-grade terrorist threat emerged. In April 2004 there were unconfirmed reports of a threat and interdicted attack on the visiting cruise ship *Queen Elizabeth II*.

As it turned out, efforts to radicalize South Africa's Muslim communities in the post-apartheid period were unsuccessful. South Africa has a long Islamic tradition, beginning during the second half of the seventeenth century with the arrival of Sheikh Yusuf, a prominent Malay cleric who became Islam's guidepost during the Dutch colonial period. His tomb remains the most important of the Five Shrines of Islam that encircles the Cape Peninsula, and his teachings of moderation and self-reliance sustained the community during the difficult days of oppression and apartheid. Around the turn of the twentieth century, data suggest, the Cape Muslim community was second only to Jawa (Java), Indonesia, in the proportion of the population that made the required pilgrimage to Mecca (de Blij, 1969). No African country lies farther from the Islamic Front than South Africa, and none matches the unique combination of Islamic influences on display in the realm's most important state.

TERRORISM'S GEOGRAPHIC INCUBATORS

When, in the aftermath of 9/11, scholars and other observers proclaimed that the world would never be the same again, their reaction was understandable. The "most successful special operation in the history of humanity," as al-Qaeda's propaganda office proclaimed it, was a culmination of two decades of growing Islamic terrorism that already overshadowed all other forms of the phenomenon (when news of the 1995 Oklahoma City terror attack first broke, the reaction of many of the same observers was that this must be another instance of it). It seemed to presage an era of ever-increasing insecurity and massive disruptions of life. "The opening salvo in Huntington's Clash of Civilizations," said more than one pundit about 9/11.

Certainly 9/11 caused major changes, but some of them were self-inflicted in the form of excessive security regulations, border delays, and other costly and disproportionate responses described as "overblown" by more than one analyst (Mueller, 2009). The war in Iraq was one of these self-inflicted changes, its substantive as well as psychological price beyond measure. But what the experts had in mind was a reign of terror that would match the challenges posed by China's rise to superpower status and the hazards presented by climate change. More than ten years later, that judgment does not seem to hold.

Successful terrorist attacks have their impact in the form of temporary disruptions and emotional agonies, but in terms of changing the world, they have virtually none. The single natural calamity of the 2004 tsunami killed ten times more people than all the terrorist attacks of the past two decades combined. The real risk lies in the acquisition of nuclear weapons by terrorist organizations with the capacity to use them— or by rogue states with a history of terrorist actions. A first incident of that kind would be the real world-changer, but that moment has not arrived.

None of this implies that the pursuit of terrorist organizations should not be a high priority—but the focus should be on weapons of mass destruction, not small-scale operators or their tribal protectors. This raises the question of geography—the *where* question. It is clear that certain political-geographical circumstances and particular natural environments create opportunities for terrorist activities. In the first place, failed or seriously malfunctioning states provide unique opportunities for infiltration because their disorder permits terrorist cells to organize and operate beyond the control of authorities, or even with their connivance as was the case in Taliban-ruled Afghanistan. Today, several states near or along the Islamic Front are in this category, prominently Somalia but also Yemen and Eritrea as well as Sudan. Yemen is Usama bin Laden's ancestral home and a base for AQAP; Somalia has become a hotbed of terrorism currently (2011) driven by al-Shabaab. In the Middle East, Lebanon's disorder and accommodation of Hizbullah remains a concern. In the Western Hemisphere, Venezuela's malfunction has become a worry; several years ago there were intelligence reports of suspicious activity on one of its Caribbean islands, Margarita Island.

Secondly, terrorist cells find refuge in remote, rugged, rural environs where they can blend in with the local population while operating their networks and preparing their attacks, especially when those environs adjoin suitable targets. The mountainous, forested eastern periphery where Afghanistan adjoins Pakistan's Waziristan (the Federally Administered Tribal Areas or FATA) is the prime example, the place where both Usama bin Laden and his Taliban ally Mullah Omar sought and found refuge when the United States commenced its intervention after 9/11. Today the remote interior of Yemen is a comparable base for AQAP,

THE "ARAB SPRING":
GEOGRAPHY TRANSCENDED

During the Indochina War, much debate in political geography centered on the so-called Domino Theory. A communist victory in Vietnam, according to this notion, would threaten and eventually topple governments in neighboring Southeast Asian states, a contagion that would threaten the anti-communist containment policy that lay at the heart of American Cold-War strategy.

While the Domino Theory may not deserve to be called "theory"—it has more hypothetical than theoretical qualities—it was never difficult to cite examples of its apparent validity, from Southeast Asia to Africa. Its simplest definition held that destabilization from any cause in one country would result in the collapse of order in a neighboring country, triggering a chain of events in a series of contiguous states. The key was a "spillover" effect that crosses borders and infects neighbors—the falling dominoes.

Today, a movement called the "Arab Spring" (or "Arab Awakening") sweeps across the geographic realm demarcated by the Islamic Front, but the dynamics are far different. It originated in Tunisia, wedged between Algeria and Libya on Africa's Mediterranean coast. When the first

which planned and launched several sophisticated although unsuccessful assaults on Western targets from this redoubt.

And thirdly, Islamic terrorists find havens in large, chaotic cities because they provide cover and anonymity as well as amenities such as phones, banks, mosques, money-laundering businesses and, of course, targets. Pakistan's coastal city of Karachi, always a violent place, has become a cauldron of terrorist activity. Baghdad in 2010 suffered the highest number of terrorist attacks of any city in the world. Peshawar in 2010 had more than ten times as many attacks as it did in 2000. The juxtaposition of a large city and a topographically or environmentally remote refuge constitutes an ideal situation: southern Somalia's proximity to Nairobi, Pakistan's Peshawar near the Afghanistan border, Buenos Aires and the Triple Frontier. Indeed, Usama bin Laden ultimately found refuge not in mountain caves but in a villa not far from Pakistan's capital Islamabad. Indeed, a group of ecosystem geographers at UCLA, led by T. Gillespie and J. Agnew, along with a class of undergraduates, wrote a massive and deadly protests erupted there in mid-December 2010 and the corrupt regime was toppled a month later, it was easy to envisage a domino effect in always-tense Algeria and autocratic Libya. But no such thing happened, at least not immediately. The "Arab Spring" touched Jordan, then infected Egypt (January 25, 2011), Yemen (January 27), Bahrain (February 14), Libya (February 16) and Syria (March 15). Look at the map, and it is clear that something other than proximity was driving the process.

That other factor, of course, was the Internet. The involvement of millions of protesters and their organized and largely orderly demonstrations resulted from the capacities of Facebook and Twitter and other connective options to appeal to the younger generation that continues to drive the movement. The Internet rendered the Domino Theory irrelevant, achieving what the domino effect could not: regionally disseminate public anger and civic energy and facilitate their coordinated expression.

In early 2012 the "Arab Spring" saw mixed results, Cairo's Tahrir Square the scene of renewed protests and casualties. But even if Tahrir Square becomes the Arab world's Tienanmen, it will forever symbolize the power of technology to surmount the barriers of geography.

report in 2009 including a probabilistic model predicting that bin Laden would be found living "in a city less than 300 kilometers (190 mi) from his last known location in Tora Bora, a region that included Abbottabad, Pakistan, where he was killed last night" (Reardon, 2011).

Bin Laden's elimination and the killing of other terrorist leaders has emboldened some Western officials to assert that the Islamic terrorist movement has been decapitated and disabled. That is likely to be a premature assertion: from the Philippines to France and from Russia to Nigeria, the world affords countless opportunities for terrorist operations, and numerous motivations to spur them on. But their efforts have not precipitated a clash of civilizations, nor have they changed the world the way al-Qaeda desired.

8

RED STAR RISING: CHINA'S GEOPOLITICAL GAUNTLET

NAPOLEON, WHO KNEW A THING OR TWO ABOUT EMPIRES AND IMPE-
rialism, is supposed to have remarked that China was a giant asleep, and that whoever woke it up would regret doing so. In the two centuries that followed, European colonial powers shook China's lethargic rulers, Japanese armies jolted the Chinese heartland, and Soviet ideologues got in bed with their Maoist counterparts. But China outlasted the colonialists, ousted the Japanese, and outdid their Stalinist advisors' communist fervor, all without entering the global stage. Nor did the Chinese retaliate against the Europeans, even allowing the British to reassert themselves in postwar Hong Kong. No state-sponsored violence was directed against Japan, whose dreadful wartime atrocities rouse Chinese passions to this day. And the Soviet advisers were simply sent home, not punished for their despised "revisionism." All the while, China was not a member of the United Nations, Chinese goods were not seen on international markets, and China slumbered on in self-imposed isolation. Convulsive nightmares such as the Great Leap Forward and the Great Proletarian Cultural Revolution, costing tens of millions of lives, periodically stirred China but had no effect on the outside world. Famines and natural disasters had little impact on population numbers: overwhelmingly rural China grew exponentially, contributing about one-quarter of the global yield of the twentieth-century population explosion.

But then, on a fateful day in 1971, President Nixon announced that he had sent an emissary to the Chinese capital to make arrangements for a visit to meet with the country's premier, Zhou Enlai, and with its Communist Party Chairman, Mao Zedong. To most Americans it seemed inconceivable that an American president would sit at a table with the rulers of the world's largest and most virulently communist nation, one whose army had killed many American soldiers during the Korean War just two decades earlier. Yet even before the visit took place, the United States announced that it would support action at the UN General Assembly in favor of seating the People's Republic of China. Although the United States also proclaimed that it would oppose the ouster of Taiwan from the UN, the resolution that admitted China simultaneously stipulated the expulsion of the Taiwanese delegation. Clearly China was about to enter the global stage, with incalculable consequences. When Nixon arrived in Beijing on February 21, 1972, it was not difficult to hear the echo of Napoleon's prescient words.

TECTONIC INTERVENTION

China in the 1970s still was embroiled in the Great Proletarian Cultural Revolution, started by Mao and his clique in 1966. Fearful that Maoist communism had been contaminated by Soviet "deviationism" and worried about his own stature as its revolutionary architect, Mao unleashed a campaign against what he viewed as an emerging elitism in Chinese society. He mobilized young people living in cities and towns into cadres known as Red Guards and ordered them to attack "bourgeois" elements throughout China, called on them to criticize Communist Party officials, and encouraged them to root out opponents of the system. He shut down all of China's schools, persecuted intellectuals deemed untrustworthy, and motivated Red Guards to engage in what he called a renewed "revolutionary experience."

The results were disastrous. Red Guard factions took to fighting among themselves. Anarchy, terror, and paralysis followed. Thousands of China's leading intellectuals perished, moderate leaders were purged, and senior teachers, elderly citizens, and older revolutionaries were tortured to make them confess to crimes they had not committed. As the economy suffered, food production and industrial output declined. Violence and famine killed as many as 30 million people as the Great Proletarian Cultural Revolution spun out of control.

The aftermath of all this hung over China during Nixon's visit, but it was not a topic of conversation. Neither China's nor the United States' internal affairs were discussed as a matter of protocol. After Nixon left, the waning Cultural Revolution continued to afflict the country, and as Communist Party chairman, the powerful Mao could pursue his objectives at will, constrained only by the more moderate Zhou. There is no doubt, however, that the Party's status and reputation suffered as a result of the Cultural Revolution. The arrant behavior toward elders by youngsters in a culture that for millennia had revered its senior citizens aroused widespread resentment and disdain. The resulting dislocation left few citizens unaffected. Confidence in Party leadership was at a low ebb.

What caused the unimaginable, the end of Mao's rule? Historians ascribe it to his failing health and loss of control during his final year, and to the power struggle that shook party and country in the mid-1970s. Undoubtedly the Cultural Revolution had left the state as well as the Communist Party in disarray. But geographers point to another factor that undoubtedly hastened the process. On July 28, 1976, twin earthquakes struck near the city of Tangshan, about 100 miles (160 km) east of Beijing. Tangshan, then a city of about 1 million, was almost completely destroyed. Loss of life and physical destruction also ravaged Beijing's port city, Tianjin, as well as parts of Beijing itself. The death toll, never confirmed by Chinese authorities, may have exceeded 700,000, making this the most deadly natural catastrophe of the twentieth century by far. So disorganized, incompetent, ineffective, and corrupt was the relief effort that word of it spread throughout a China already spent by a decade of Mao's Cultural Revolution. The Chairman for Life died on September 9, just six weeks after Tangshan. It probably was no coincidence.

If Nixon's opening to China caused misgivings in the West (and, to be sure, in the Soviet Union, no longer on China's ideological page), it also contributed to qualms in China itself. Mao adamantly opposed changing the political or the economic system that had made the People's Republic what it was. Zhou was in no position to change Mao's mind, but as premier he was able to promote party leaders who took a more pragmatic view of China's needs. One of those leaders was the man who would guide post-Mao China from torpor to tiger: Deng Xiaoping. In the fateful year of 1976, when power struggles and natural disasters buffeted the state, Deng readied policies that would combine continued communist dictatorship with market-driven economics. Within decades, not distant generations, China was a power to be reckoned with on the global stage. By the turn of the century it was a nuclear power, an economic heavyweight, a military juggernaut with the world's largest standing army and a growing navy, and a political force with increasing influence in its own backyard and beyond.

WAS IT INEVITABLE?

Most geographers were not surprised by China's ascent any more than they were by the Soviet Union's disintegration. The position of Eurasian states in global geopolitics had been a topic of debate for more than a century and had generated some penetrating analyses and remarkable forecasts. The first shot in this productive exchange was fired on a snowy night in January 1904 in London, when a Scottish geographer named H. J. (later Sir Halford) Mackinder delivered a lecture titled "The Geographical Pivot of History." The winter weather kept his audience rather small for a major event at the Royal Geographical Society, but from the minutes of the meeting it is clear that those present were aware that they had heard something extraordinary. When the paper was published in the *Geographical Journal* some months later, complete with a transcript of the discussion that followed, it raised a storm of reaction and for decades afterward was the most often cited article in geography (Mackinder, 1904).

Essentially, Mackinder argued that while British naval forces had made Britain the leading superpower of the time, impregnable land-based strongholds would come to dominate geopolitics. A quite exhaustive analysis of resources and risk factors led him to conclude that the Pivot Area, that is, interior Eurasia, contained the wherewithal to allow a future power to successfully challenge for world domination. This Pivot Area—he later called it the Heartland, and his idea became known as the Heartland Theory—lay in Eastern Europe and western Russia. At a time when Russia was a collapsing, overextended country losing a war against Japan on its eastern flank, Mackinder foresaw its emergence as a world power. When the Soviet Union emerged after his death in 1947 as one of the two power cores in a bipolar world, Mackinder's original article got a lot of posthumous attention.

Until the year of his death, Mackinder participated vigorously in the ongoing debate his Heartland Theory had generated. Perhaps Mackinder's most perceptive critic was a geographer named N. J. Spykman who, in the 1940s, published a book in which he compared the assets of Mackinder's Heartland with those of the Eurasian Rimland (Spykman,

1944). He probably was the first observer ever to use the term "rim" for the Eurasian periphery, foreshadowing the commonplace "Pacific Rim" to which we refer today. Spykman concluded that no matter how powerful a Soviet Union would emerge from the Second World War, its strategy should be to extend its domination into the Eurasian Rim: failing this, a Eurasian Rim power or alliance of powers would ultimately and successfully challenge for supremacy.

There is an interesting geographic order to what has happened in Eurasia over the last several thousand years. In the west, the succession of power proceeded on a latitudinal ladder, from Egypt to Crete to Greece to Rome and finally to Western Europe. Ever larger spheres of dominance attended this sequence, from the Egyptians in North Africa to the Greeks of Alexander and Persia to the Romans and their Mediterranean empire and, ultimately, to the Western Europeans and their global colonial realms. But then the succession took on a longitudinal dimension. Insular Britain was the world's first naval superpower. Mainland Germany mounted two major wars, the second for world domination. Next came the Russian (Soviet) challenge, the Heartland power at its zenith. Is it now time for the cycle that began on the western perimeter of Eurasia to reach the eastern rim?

The United States, from its ocean-moated national fortress, crucially influenced the course of events affecting all three of these powers, beginning with its obstruction of British colonialism. The Second World War and its aftermath in Korea left America on the doorstep of East Asia from Japan and Taiwan to the Philippines and Vietnam. The Cold War marked a half century of bipolar geopolitics and, when it ended in the disintegration of the U.S.S.R., left the United States disproportionately powerful in a world where, until the rise of China, there would be no clear Number Two.

POWER ON THE PERIPHERY

China's takeoff during the past 30 years unleashed its economic potential and witnessed the greatest short-term regional transformation in the history of humanity. "Pacific Rim" became bywords for reconstruction, renovation, reformation, modernization at a dizzying pace. The whole of coastal China, it seemed, was one vast construction site from Guangzhou to Dalian. Adjoining Hong Kong, the fishing village of Shenzen with its duck ponds and pig pens became a skyscrapered metropolis of 9 million in less than three decades, the fastest-growing city in world history. Across the Huangpu River from Shanghai's historic colonial avenue known as the Bund, the rural countryside was swept away and the futuristic skyline of Pudong replaced it. Gleaming new airports, high-speed intercity highways, multilane urban ringroads, state-of-the-art suspension bridges, hydroelectric projects, vast factory complexes, and modern container-port facilities reflected China's burgeoning economic power. Foreign investment from Japan, the United States, even (indirectly) from Taiwan propelled the process, and on the insatiable American market Chinese goods sold in quantities unimagined before, creating a huge trade surplus for Beijing and financial reserves undreamed of even by Deng Xiaoping himself.

With economic strength comes political clout, and by the turn of the century the outlines of a new geopolitical relationship between the United States and China were visible. After the Second World War, the Japanese had become allies of the United States; after the Korean War, South Korea achieved democracy as well as economic success. American troops remained based in both Japan and South Korea. The United States role in protecting Taiwan from Beijing's designs stopped short of stationing American forces there, but United States warships patrolled the waters between mainland and island during times of tension. The United States had military bases in the Philippines (now vacated) and Singapore, fought a war in Vietnam without significant Chinese involvement, and maintained a western-Pacific presence from Guam to Palau. As Chinese students never tire of telling me when I visit China, the United States is still on China's doorstep, but China is not on America's.

The United States, as the public learned in 2001, also conducts intelligence operations in the western Pacific, as close to China as international law allows (and undoubtedly closer). On April 1, a United States spy plane flying just outside the 12-mile limit collided with a Chinese fighter jet that was tailing it; the fighter plane went down but the spy plane, though damaged, landed on Hainan Island where the crew was briefly detained. The incident aroused much public anger in China, and although it was quickly settled its impact on public opinion in China lasted longer. This and other provocations such as the mistaken bombing of the Chinese embassy in Yugoslavia by United States warplanes during the Kosovo crisis (at fault was an outdated map: they hit the wrong address) all contributed to still another byproduct of economic success and strategic frustration: a rising tide of nationalism. I have watched and experienced this upsurge during 30 years of very nearly annual visits to China. The perceived superpower arrogance of the United States, its omnipresence in the region, its role involving Taiwan, criticism of China's human-rights practices, refuge given to dissidents, and other irritations drive nationalist sentiments expressed in newspaper editorials, letters to editors, public reactions to perceived slights, and in virulently anti-American and nationalistic best-selling books such as *China Can Say No* (Song Qiang, 1999).

I can attest to the rise of Chinese nationalism on several fronts. My geography textbooks are sold worldwide, and have been translated into Chinese as well. I receive a steady stream of critical comments from students and other readers at home and abroad, including many from Chinese students now enrolled at United States universities and colleges, about the way their countries and regions are represented. Chinese readers tend to be especially displeased by my maps: these do not clarify that Taiwan is in fact a province of the People's Republic, do not show parts of India and Kazakhstan to be Chinese territory, do not, as one angry correspondent wrote, reveal Russia's Pacific Southeast as "stolen lands," and do not assign disputed Pacific islands to China. On the other hand, a Manchu mindset seems to be as strong in Taiwan as in Beijing: my carefully nuanced narrative on Tibet (Xizang) evoked a passionate denunciation from Taipei: "The boundaries established during the Qing Dynasty are the irrevocable borders of China, and you have no right to judge our historic heritage."

Will the United States and China find themselves in a geopolitical confrontation in the decades ahead? Certainly the potential exists. When the topic arises in discussion in China, the issue of Taiwan inevitably comes to the fore. Chinese mainlanders overwhelmingly regard this as a domestic matter to be resolved by negotiation if possible but by force if necessary; American protection of Taiwan's quasi-independent status is despised and resented. Under the umbrella of American security guarantees, Taiwan has not only moved from autocracy to democracy but has begun debating the merits of independence, a prospect that provokes Chinese threats of armed intervention at any cost, even military confrontation with the United States. Former president George W. Bush, during a visit by China's leader Hu Jintao in 2002, reiterated American disapproval of any such action by Taiwan, a position sustained by the Obama administration. Meanwhile the United States sells Taiwan modern weapons. Over the issue of Taiwan, a small-scale Cold War is already in progress, and it is fraught with risk. The United States is committed to protect Taiwan against military intervention; the Chinese are placing rockets aimed at Taiwan on the mainland side of the Taiwan Strait. Even as Taiwanese investment helps drive the economic growth of China's Pacific Rim and tentative low-level discussions go on between Taipei and Beijing, political leaders engage in provocations and militaries flex their muscles. It is a recipe for trouble.

Despite encouraging signs of progress—in 2011, when the United States agreed to "modernize" Taiwan's jet fighters, China complained but did not retaliate—tempers tend to flare when the Taiwan issue comes up. After a talk a student asked me, to scattered applause, how Washington would like it if "a bunch of terrorists" took over Puerto Rico and China and prevented the United States from blockading the island and regaining control over it. During the first decade of the century, American determination to export democracy to countries that did not have it and may not want the U.S. version of it created unease in this, the world's largest—and still one-party-ruled—nation. The memory of the 1989 prodemocracy movement violently quelled on Tiananmen Square may be fading, but it is not expunged. In 2011, when the "Arab Spring" led to regime change in a number of affected countries, China joined Russia in the United Nations by voting against (and thus vetoing) action to constrain the murderous minority rule of Assad's clique in Syria. It was a vote that reflected China's own fear of destabilization, glimpsed in 1989 and revived by the events that roiled Egypt and Libya in the wake of the initial protests.

Competition between the United States and China, furthermore, is not confined to the political arena despite the commercial and fiscal embrace in which the two countries find themselves. China gets a great deal of blame for the outflow of American jobs to foreign places and for keeping the value of its currency lower than it should be; listen to American media commentators describing the economic disadvantage in which the United States finds itself and you would think that the new Cold War is not only here but already lost. By 2010 China had become the world's second-largest economy, and projections suggested that China's economy would surpass that of the United States somewhere between 2020 and 2030. But China has its own problems: the transition from a dominantly export economy to an economy driven by the spending of domestic consumers will not come easily.

Meanwhile, a far more consequential byproduct of China's economic growth is the country's escalating demand for raw materials and energy resources. From copper and coal to "rare earths" and oil, China's needs are gargantuan. Not many years ago, China was able to meet its energy needs from domestic sources, mainly its oil reserves in the far west and from coal deposits in several parts of the country. Today, China must import oil, and in fast-growing quantities; its demand is having a major impact on global energy trade. The most direct supply lines are from the Caspian Sea area and from Russia, and, fortunately for the Chinese, only one (though territorially huge) country lies between China and the Caspian Sea: Kazakhstan. Agreements between Beijing and Astana have allowed for the start of construction of pipelines from the Tengiz oil reserve in the Caspian Basin to Urümqi in western China. Meanwhile, the Chinese are negotiating with the Russians to extend Russia's pipeline network into China's northeast, but here the Chinese are meeting competition from the Japanese, who view Russia's eastern oil reserves as vital to their future. Most of Japan's enormous demand for oil and natural gas is met by long-range tankers, but disruption of supply lines is a growing risk in this era of political instability and terrorist threats in the overseas source areas. Access to supplies from the comparatively nearby Russian oil fields is crucial to Japan's economic prospects. The Japanese want Russia to provide oil via a coastal terminal across from Japan. However this process plays out, China's emergence as a large and growing consumer of a nonrenewable global resource of which the United States is by far the world's largest consumer and of which America has only limited domestic reserves will affect relations between the two states as time goes on. The question is not whether these relations will become more difficult; rather, it is how to prepare for and manage the difficulties when they arise.

CHINA'S SURPRISING GEOGRAPHY

One way to mitigate the problems ahead is to comprehend them better. In the introductory chapter I argued that America's rather sudden ascent to the status of sole superpower confers on Americans a responsibility to further internationalize their outlook, to be better informed about the countries and societies on which the United States has such an enormous impact—and that one good way to go about this is to study global geography. Surveys indicate that Americans' geographic knowledge of China remains dim, despite our deep involvement in the burgeoning of

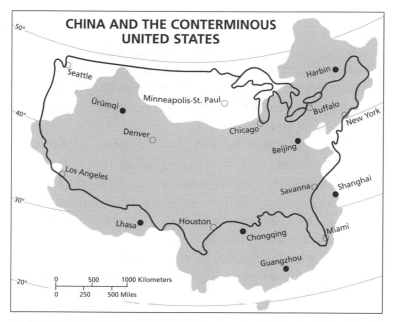

Fig. 8-1. China and the United States in their respective latitudinal positions, showing not only comparative size and distance but also latitudinal extent. China extends both farther north and farther south than the contiguous United States, but the east-west dimensions are almost identical.

its economy. In my own experience, though, we tend to be fascinated and often surprised by even the basics of China's regional geography.

Take just one example: the map of the conterminous United States superimposed on a map of China (Fig. 8-1). This map is not only to scale, it also puts the two countries at their appropriate latitudes. Many Americans think that China is much larger than the United States, given its huge population. But in fact, including Alaska, the United States is slightly larger (9,629,047 against 9,573,945 square kilometers). In the east, though, China extends farther north even than the latitude of northern Maine, and farther south than southern Florida. China's northeastern provinces have bitterly cold winters and short summers, and a sliver of Russian territory landlocks them from the Pacific Ocean. In the south, China is truly tropical: the island of Hainan lies in the general latitude of the Greater Antilles. Hong Kong and Guangzhou lie well south of Miami.

Westward, both countries become more mountainous and drier, although more gradually in the United States than in China. In the United States, relatively mild conditions prevail into the Great Plains; in China, forbidding, ice-capped mountains in the south and vast deserts in the north dominate a physical landscape with comparatively few oases of cultural life. Although it is about as far from Shanghai to Urümqi as it is from Savannah to Los Angeles, there is no Los Angeles in China—nor is there a west coast. China's far west abuts the steppes of Mongolia and the mountains of Turkestan, historic frontiers of contact between Chinese and non-Chinese in remote inner Asia.

Which brings us to what is surely the best-known aspect of China's geography: the country's huge human population, now exceeding 1.3 billion, more than one-fifth of all humanity and, despite a declining growth rate, still adding nearly 7 million annually. With such numbers, you would imagine that every part of China is densely peopled, but in fact the opposite is true. Ninety-five percent of China's population is concentrated in the eastern one-third of its territory, and even in this eastern third there are vast, sparsely populated areas. Natural landscapes have a lot to do with this—people tend not to agglomerate in icy mountains or in arid deserts—but the distribution is also a matter of historical geography. China's origins lie in the fertile farmlands of its great river basins, the Huang (Yellow) and Yangzi-Chang in the middle, the Liao-Songhua in the north, and the Xi (West)-Pearl in the south. Here, more than 5,000 years ago, arose the communities that would one day be collocated into a cohesive state, the core of an empire that eventually reached far into Asia in all directions. To this day, and despite all the dramatic events along its Pacific Rim, China numerically remains a dominantly rural society. Behind the façades of Hong Kong and Pudong lie vistas of teeming rural countrysides where farmers by the millions bend to the soil each day as they have for millennia. Their produce, taxes, and tribute built the state; they were persecuted, conscripted, expropriated, dislocated, starved. But in China's great river basins they survived, laying the foundations of a demographic map that has changed little over thousands of years.

Today their world is changing, but old habits die hard. Pacific-Rim China has urbanized faster and more pervasively than any other part of the world, and the numbers are simply amazing. China was just 18 percent urbanized in 1978, 26 percent in 1990, and 50 percent in 2012. Between them, the largest city (Shanghai, over 17 million) and the capital, Beijing (nearly 13 million) have more inhabitants than a majority of countries of the world. *One hundred* Chinese cities have more than 1 million people; by 2025 there may be as many as 200. Yet behind all these spectacular numbers lie troubling realities. While wealth is concentrating in this New China, hundreds of millions of citizens in the rural provinces continue to live in poverty. Would-be workers from the villages walked and biked by the millions toward the jobs in the shadows of all those new skyscrapers, but eventually the authorities were forced to stem the migration and reverse it. In the headlong rush to modernize, China's planners victimized the powerless farmers once again. Back home, they found the tentacles of Communist-Party power intruding into their lives as well-connected communist capitalists brought the market economy to their villages in the form of factories and often-unwanted public projects— even as east-west income inequalities continued to widen.

Politically powerless under conditions of endemic corruption, subject to land expropriation as well as pollution of air and water from new mines and factories in their midst, and disconnected from the forward rush of the New China, the country's farmers—still the majority—have difficult lives and suffer risk and uncertainty. Improving this situation, reducing the income gap and giving villagers a stronger political voice will be China's great domestic challenge as its international stature grows. Failure could endanger the entire national enterprise.

China probably is the oldest continuous civilization on the planet (Egypt might contest the claim), and its sequence of dynastic rule began in the area where the Huang (Yellow) and Wei Rivers meet, not far from the modern capital on the North China Plain. The country alternately fragmented and reunified, absorbed and assimilated outsiders and conquered neighbors, but reached an apogee during the Han Dynasty (206 BC–AD 220), when the Roman Empire thrived at the other end of Eurasia, when the city of Xian was the "Rome of China," and when the Silk Road was a busy trade route. To this day, Chinese call themselves the "People of Han." The Ming Dynasty (1368–1644) was another formative dynasty, but China reached its greatest imperial dimensions during the final traditional dynasty, the Qing (Manchu), which commenced in 1644 and ended in chaos in 1911. The Qing rulers conquered much of Indochina, Myanmar, Tibet (Xizang), Xinjiang, Kazakhstan, Mongolia, eastern Siberia, the Korean Peninsula, and the islands of Sakhalin and Taiwan (Fig. 8-2). Current claims to Taiwan and Tibet and latent claims to Russia's Far East and India's northeast are based on these imperial boundaries, but disaster was in the offing. The nineteenth century was calamitous for a China weakened by overextension, European colonial invasion, and Japanese intervention. British merchants undercut China's economy;

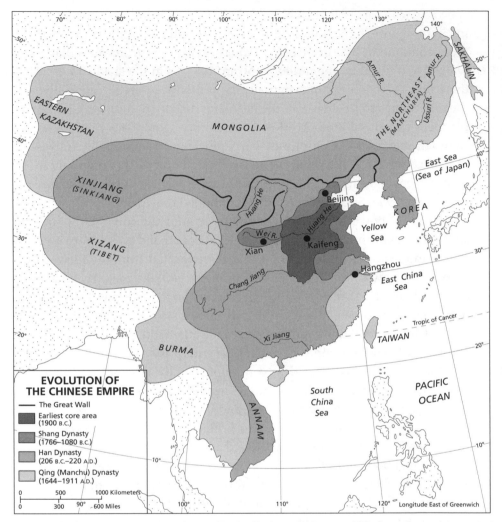

Fig. 8-2. Evolution and growth of the Chinese Empire. Versions of this map, which shows the maximum extent of Qing-Dynasty China, often accompany comments by Chinese readers of my books to justify Beijing's control over Tibet (Xizang), Mongolia, and even parts of Kazakhstan and Russia.

opium flooded into China from India and destroyed the very fabric of Chinese society. When the Qing rulers tried to stop this British import, they were defeated in what is called the First Opium War (1839–1842) and the breakdown of Chinese sovereignty was under way. When the Qing Dynasty collapsed in 1911, the empire's territorial losses were already enormous, but worse was to follow as nationalists, communists, and the Japanese fought a three-way war that ravaged the country.

Following the defeat of Japan in the Second World War, China's civil war resumed and Mao Zedong's communists drove Chiang Kai-shek's nationalists from the mainland to Taiwan, where they overpowered the locals and set up their own "Republic of China." In Beijing on October 1, 1949, Mao Zedong proclaimed the birth of the People's Republic of China.

CHANGING THE MAP

China's twentieth-century convulsions were not unlike those of pre-Yuan Dynasty times: the chaotic end of a dynasty amid a breakdown of order, regional fragmentation and civil war, interference from outsiders (in this case not the Mongols or the Manchus but the Europeans and the Japanese), decades of strife ending when a powerful new leader took control. Indeed, after only a few years of Mao's rule scholars began to draw parallels between earlier dynasties and this new communist one. Mao might not have offspring waiting in the wings, but the intrigues and conspiracies marking communist rule and succession were worthy of the dynastic palaces of old.

Mao inherited a weakened China that had lost Taiwan, had British and Portuguese colonists on its doorstep in Hong Kong and Macau, had been forced to accept unfavorable boundary delimitations by more powerful neighbors, and had only nominal influence in far-flung reaches of the state. All these disadvantages would be dealt with eventually, but first the communist regime wanted to address the fate of the millions of landless people and serfs whose lives were indescribably miserable. For all its historic grandeur and cultural achievements, China remained a country with rural areas where floods, famines, and disease could decimate entire populations without any help from the state, where local lords could (and often did) repress the people with impunity, where children were sold and brides were bought. To deal with these horrors, Mao mobilized virtually every able-bodied citizen. Land was taken from the wealthy, even from the small subsistence farmer. Farms were collectivized, dams and levees were built with the bare hands of thousands. Food distribution was improved, public health enhanced, child labor curtailed. But in the process, China's communist planners also made terrible mistakes. The so-called Great Leap Forward, started in 1958 as the final transformation of China's social life, required the reorganization of the peasantry into communal brigades to combine local industrialization with more productive farming. Peasants were forced into collectivized villages, families were torn apart, and the historic rhythm of agriculture was severely disrupted. Between 40 and 50 million people had died of starvation and mistreatment by the time the Great Leap Forward was abandoned in 1962 (Vogel, 2011).

Meanwhile Mao had moved to reassert Han Chinese influence in the country's distant frontiers. Some major losses of the late Qing period could not be reversed, but others could be prevented. Chinese armies entered mountainous Tibet (Xizang) soon after Mao took control in Beijing, and when Tibetans rose against the Chinese in 1959, their revolt was ruthlessly crushed and the Buddhist Tibetans' supreme leader, the Dalai Lama, was ousted. Tibet's historic, fortresslike monasteries, around which Tibetan life was organized, were emptied and much of the fragile society's cultural heritage was destroyed. In desert Xinjiang, peopled by Uyghurs and other Muslim groups with ties to Turkestan, Mao began a program of vigorous Sinicization, the imposition of Han rules and norms. In 1950, Chinese made up only about 5 percent of Xinjiang's population. Fifty years later, the proportion approached 50 percent (of more than 20 million). Here, too, the Chinese faced opposition from its colonized minorities, opposition that more recently has become enmeshed in the global Islamic-terrorism issue.

Two other Maoist initiatives stand out in any assessment of his impact on China's social geography. The first is demographic. Mao often referred to any control of population growth as a capitalist plot to weaken communist societies, and, like his cohorts in Moscow, he encouraged mothers to have numerous children (in the 1960s, Soviet women bearing ten children or more were designated "Heroines of the State"). As a result, following the end of the Great Leap Forward, population growth in China during the 1960s and 1970s was as high as 3 percent, and while the official figures are unreliable, China was adding as many as 20 million per year, rushing toward 1 billion. When Deng Xiaoping's pragmatists took over following Mao's death, one of their priorities was to get this spiral under control, and they instituted their infamous "One Child Only" policy to achieve this. Today, China's official growth rate is 0.5 percent, which, on a base of more than 1.36 billion, still results in an annual increase of about 7 million.

The second Maoist initiative was to try to erase the legacies of China's most influential philosopher and teacher, Confucius (Kongfuzi or Kongzi in modern Chinese). Born in 551 BC, Kongfuzi's teaching dominated Chinese life for over 20 centuries. Appalled at the suffering of ordinary people, he encouraged the poor, taught the indigent, questioned the divine ancestries of aristocratic rulers and poured forth a mass of philosophical writing that became guiding principles during China's formative Han Dynasty (206 BC–AD 220). Over time, a mass of Kongfuzi literature evolved, much of which he never wrote but might have approved. At the heart of it lay the Confucian Classics, 13 texts that became the basis for education in China for 2,000 years. From governance to morality and from law to religion, the Classics were Chinese civilization's guide. State civil service examinations, through which poor as well as privileged could gain access to positions of influence, were based on the Classics.

Kongfuzi championed the family as the foundation of Chinese culture, which was incompatible with basics of the Great Leap Forward. He also prescribed a respect for the aged that was a hallmark of Chinese society, but contradicted tenets of the Great Proletarian Cultural Revolution. The communist regime attacked Kongfuzi on all fronts, but Beijing miscalculated. It proved impossible to eradicate two millennia of cultural conditioning in a few decades. After Mao's death, public interest in Kongfuzi surged despite the modernization of education that had been started by the communists, and when you walk into a bookstore in China today you will see shelves sagging under the weight of the Analects and other Confucian writings. Westerners are paying renewed attention too. The last British governor of Hong Kong, Christopher Patten, begins all but one chapter of his reflective book with a quotation from the Analects (Patten, 1998). The spirit of Kongfuzi will pervade physical and mental landscapes in China for generations to come, and understanding this is a key to interacting with the culture.

THE MODERN MAP OF THE CHINESE EMPIRE

Make no mistake: China remains a modern-day empire. When the Soviet Union collapsed, numerous scholars (to my knowledge, none of them geographers) wrote treatises about the fall of the "last empire." More recently one of my colleagues wrote a book entitled *The New Imperialism* suggesting that the United States is the world's last-gasp empire (Harvey, 2004). But the enduring empire is China. Even after the territorial losses sustained during and following the disintegration of the Qing Dynasty, Beijing was left with an imperial realm extending from the Korean border to Turkestan and from Mongol steppes to Tibetan valleys. First the communists, then Deng Xiaoping's pragmatists organized this vast contiguous empire with its numerous minority peoples into a spatial framework that would accommodate, at least in principle, their respective objectives.

China's political map has changed repeatedly in recent years, and it will undoubtedly change again. Some of the changes made by the post-Mao administrations reflect a desire in the capital to directly control critical parts of the country. Other changes have been made to facilitate the wedding of communist politics to capitalist economics. Additional modifications have come with the changing status of former European colonies now reabsorbed into the motherland. Should the Taiwan issue be unexpectedly resolved, the map will change again. Meanwhile, new or newly prominent names make their appearance: Shenzhen, Pudong, Xianggang (for Hong Kong). China's geography is a work in progress.

As Figure 8-3 underscores, China's modern political framework has four layers of administration. At the top are four central-government-controlled municipalities or *shis*, really the crucibles of Chinese culture and power: Beijing, the capital; Tianjin, its neighboring port; Shanghai, at the mouth of the Yangzi; and Chongqing, upstream near the great Three Gorges Dam. On the map, Chongqing looks much the largest, and territorially it is: herein lies a geographic tale. The city of Chongqing is no giant, but its hinterland lies in the densely peopled eastern part of the Sichuan Basin, a cluster of more than 100 million people clearly visible on any population map of China, even a smaller-scale one. When it was decided to make Chongqing a shi, its municipal boundary was delimited far into its hinterland, adding as many as 20 million people to the "municipal" population and putting them under Beijing's direct control. That made Chongqing, suddenly, the "world's largest city," as its brochures proudly proclaim. But take this with a grain of salt. The suburbs do not generate much commuter traffic.

Next come China's 22 provinces. Here is another similarity to the United States: China's smaller provinces lie in its east and northeast, and its larger ones, territorially, lie in the west. But the similarity ends here. Look at the populations of these "small" eastern provinces, and any image of comparative smallness disappears. The four provinces facing the Bo Hai (Gulf) contain more people than the entire United States. Thus it is worth noting the locations and populations of some of China's prominent provinces (everyone in coastal China seems to know about New York, Texas, and California). Even after "losing" 20 million people to Chongqing city, Sichuan is a province with nearly 90 million people,

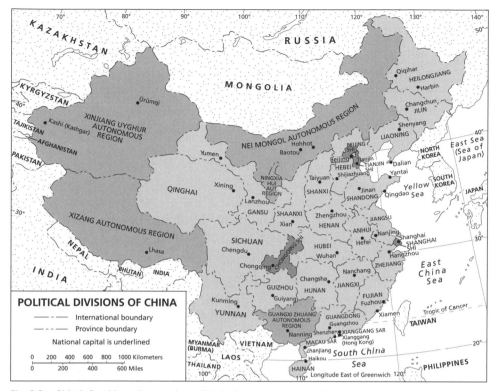

Fig. 8-3. China's four-hierarchy, complex political-administrative map. The so-called Autonomous Regions are undergoing Hanification; they are autonomous In name only although minorities are exempt from certain regulations that apply to Han Chinese. The province of Qinghai contains a substantial Tibetan-Buddhist minority.

roughly the population of Germany, astride the river that begins its course named the Chang and ends it (near Shanghai) as the Yangzi. One of China's agricultural breadbaskets and now the site of the world's largest hydroelectric project, linked to the coast by a major waterway and flanked by one of China's four shis, Sichuan lies at the head of a wedge that is carrying the economic successes of the Pacific Rim into the country's interior. This momentous development heralds the transformation of China's interior, the first step in reducing the regional disparities between the booming coast and the lagging center.

Two adjacent provinces, landlocked Henan and peninsular Shandong, each have populations in Sichuan's range, approaching 100 million in 2012 (that's more than California, Texas, New York, and Florida combined!). Along with neighboring Hebei Province (70 million) they anchor China's core area, including the capital and the key port of Tianjin.

But the provinces whose names are synonymous with the economic rise of China's Pacific Rim are coastal Jiangsu and Zhejiang on either side of Shanghai; Fujian directly opposite Taiwan and for centuries the source of "overseas Chinese" who emigrated to Southeast Asia and whose wealth returned to propel the modern Chinese economy; and Guangdong in the south, where the Pearl River Estuary is evolving into one of the world's greatest urban-industrial complexes incorporating Hong Kong, Shenzhen, Guangdong, Zhuhai, and Macau.

It was Deng Xiaoping's notion to apply his new economic policies in this coastal zone, where market economics and communist politics would coexist without contaminating the rest of the country. Accordingly, the regime introduced a complicated but effective system of so-called Special Economic Zones (SEZs), which in effect were a series of port cities and coastal areas where foreign technologies and investments were welcomed and where investors were offered capitalist-style incentives. Low-wage labor could be hired, taxes were low, leases were simple, and products could be sold on foreign as well as domestic markets. Even Taiwanese enterprises could operate here, though under certain restrictions. The results were spectacular. American and other foreign companies flocked to China's Pacific Rim, jobs by the millions were created, goods poured from newly built factories, and long-dormant coastal cities like Shanghai, Xiamen, Shantou, and even Hainan island's Haikou were transformed by modern office towers, highrise apartments, business parks, and factory complexes. Shenzhen, the SEZ adjacent to Hong Kong, led the way in terms of growth and production, but the whole frontage on the Pearl River Estuary witnessed unprecedented urbanization, industrialization, and modernization. Visitors who have not seen China since the 1980s gasp at this landscape of economic power, its ultramodern bridges, tunnels, highways, trains, airports. It is worth remembering that much of this is happening in China's tropics. Some economists persist in equating tropical environments with "underdevelopment."

This momentous regional growth is not what Mao Zedong anticipated or, if certain Chinese historical geographers are correct, would have preferred. Mao wanted to make China's Northeast (the historic home of the Manchus who had led the Qing Dynasty) the industrial heartland of the People's Republic. The Japanese had occupied this area in the 1930s, and had set up a puppet state called Manchukuo there. China got most of it back, though the Russians held on to a large sector taken from the Qings, and Mao planned to integrate its three provinces tightly into the Chinese command economy. The Northeast had large coal reserves and sizeable oil fields as well as a wide range of minerals including iron, aluminum, manganese, copper, molybdenum, magnesite, lead, and zinc, extensive forests stood ready for exploitation on the slopes of the far north, the Japanese had laid out a substantial infrastructure, the area lay close to the capital, and there was plentiful labor.

The southernmost Northeastern province, Liaoning, was best endowed in every respect and the basin of the Liao River was agriculturally productive as well. Its port, Dalian, was the outlet for the whole Northeastern region. By the end of the 1950s, huge state investments in mines, factories, and transport infrastructure had made such cities as Anshan, Fushun, Shenyang, and Harbin industrial giants, and the provinces of Liaoning, Jilin, and Heilongjiang drew hundreds of thousands of workers to their manufacturing plants.

But virtually nothing produced here, from locomotives to toys and from appliances to furniture, reached, or could compete on, world markets. China's communist planners, like their counterparts in the Soviet Union, assigned production quotas to mines and factories whatever the cost, and even if a locomotive could have been bought at half the price in Canada, the manufacture of railroad equipment continued in Liaoning. China was a closed society, almost a closed system; factories produced as a matter of state policy and workers were secure.

When China's economic transformation took off, the Northeast was left behind. The reformers who took over in Beijing after Mao's demise and who opened China's economic potential to the world had trouble combining market principles with state support for loss-making industries. Traveling from the Beijing-Tianjin area northward around 1990 you would have seen rustbelt scenes reminiscent of parts of Atlantic and Northeast States in the U.S., only at a much larger scale. As government subsidies declined, a few factories managed to adjust and compete in the new economic order. But most were unable to do so, and plant after plant closed, leaving their workers unemployed and, having expected job security and retirement, often in desperate straits. Beijing tried a combination of social programs and continuing subsidies, but this was a huge and intractable problem. You might ask why those workers didn't just move south, into the fast-developing Pacific Rim region, but there they would have to compete for low-paying jobs against the eager villagers from the west—which was not something they were prepared for. Instead, thousands of northerners crossed the border with Russia as traders and construction workers (until the Soviet Union collapsed and construction projects halted).

In the early 2000s things began to look up, first in and around the port city of Dalian, then farther north. As labor costs rose in the booming south, the low wage expectations of Northeastern workers attracted the attention of companies based in such places as Guangzhou and Shenzhen. Foxconn, a subsidiary of Taiwan electronics giant Hon Hai, moved operations from Guangdong province in the south to Henan in the Northeast in 2010, and other technology companies followed suit. The Pacific Rim "miracle" had finally reached the Northeast.

China's meteoric economic rise had produced some remarkable data. In 2011 China had more than 130 billionaires (in U.S. dollars) and an estimated 900,000 millionaires. But one in six of its more than 1.3 billion people—for a total of 225 million—lives on the equivalent of $1.25 per day or $450 per year. For them, the miracle is a mirage.

The third political-administrative level is a relatively new one: the so-called Special Administrative Region (SAR). This category was established to recognize the particular circumstances of Hong Kong's transfer from British dependency to Chinese administration in 1997, and in anticipation of Beijing's takeover of Macau from Portugal in 1999 (a few optimistic observers see it as facilitating the reabsorption of Taiwan at some future time). The notion of "One Country, Two Systems," was much in vogue during the week of Hong Kong's transfer. Hong Kong would become part of China, but it would retain trappings of democracy not available to citizens in China itself. As a member of ABC's television staff I interviewed several British as well as Chinese officials in Hong Kong as the day approached, and it was clear to me that the phrase held different meanings for the two sides. British representatives envisaged a durable democratic system in the ex-colony (which, of course, the British themselves did not nurture until the very end of their tenure, when the so-called Basic Law gave Hong Kong its own constitution). Eventually, they said, the territory's chief executive, at first appointed by Beijing, would be elected directly by Hong Kong's voters. Chinese officials seemed to see it as facilitating the transition but not as a long-term prospect. People would see, they said, that there were more important things in life than democratic politics. The elections of September 12, 2004, appeared to confirm their views. Although the Beijing regime had made it clear that no direct election of the SAR's chief executive was in prospect, and pro-Beijing activists in Hong Kong had tried crudely to tarnish the reputations of prodemocracy candidates, the vote went in favor of pro-Beijing parties and suggested that, indeed, many voters were more concerned about economic conditions than political freedom.

Look at Figure 8-3 one more time, and you will see that vast segments of the Chinese empire are categorized as Autonomous Regions (AR), the fourth level in the country's administrative hierarchy. The Soviet Union had its supposedly autonomous Soviet Socialist Republics: China has its supposedly Autonomous Regions. The terminology may reflect a history of good intentions, but the reality is rather different. Established originally to recognize non-Han minorities living under Chinese control, the five ARs have been exempt from certain national laws (for example, minorities were not subject to the One Child Only regulations instituted under Deng Xiaoping), but they have no autonomy in the practical sense of that word. Xizang (Tibet), formally annexed in 1965, functions as a colony of China, and even though Beijing's harsh rule was relaxed after 1976 and much looted treasure was returned, no expressions favoring independence are tolerated. The Xinjiang-Uyghur AR, where China has still-important oil reserves and where its nuclear weapons, rocket, and space programs are located, is being rapidly Sinicized even as the Uyghurs and other Muslim minorities are tightly controlled. A small pro-independence movement has engaged in occasional acts of sabotage, to which China has responded forcefully; a new railroad to the traditional Muslim city of Kashi (Kashgar) in the far west symbolizes Beijing's determination to confirm its mastery in this Autonomous Region. In the Nei Mongol AR, too, massive immigration has changed the ethnic balance. Originally established to recognize the rights of the several million Mongols who live on the Chinese side of the border with Mongolia, its cultural landscape is now dominantly Chinese. Near the Mongolian border, Mongols still traverse the steppes with their tents and goat herds, but elsewhere this is now a region of farm villages and industrial towns—and Chinese outnumber the Mongols by more than four to one.

As the map shows, there are two other Autonomous regions in China, the Ningxia Hui AR adjacent to the Nei Mongol AR and the Guangxi Zhuang AR in the far south, between the Vietnamese border and Guangdong Province. The Ningxia Hui AR is of interest because it is the historic base of the Hui, who are descendants of Chinese who were converted to Islam when it penetrated western China during the seventh and eighth centuries. The Guangxi Zhuang AR, with nearly 50 million inhabitants, is by far the most populous of the five ARs. The first half of its tongue-twister name refers to the Chinese of the east, and the second half to the Zhuang, a people with ethnic links to the Thai, who are dominant in the interior west. Numerous scattered minorities continue to speak their traditional languages here, including the Yao, Miao, and Dong.

To be sure, this girdle of Autonomous Regions around Han China does not begin to reflect the range or the distribution of minorities throughout the empire. From the Koreans in Jilin to the Li and Miao on Hainan Island, and from the Yi and Dai of Yunnan to the Tibetans of Qinghai, minorities representing about 50 ethnolinguistic groups, most with historic ties across China's borders, live under Beijing's rule. For minorities outside the officially designated ARs, reserves have been set aside in which minority rights are recognized, but often these are also poor and erosion-afflicted domains. On the one hand, the historic strength of Chinese traditions and more recent economic developments have acculturated millions of minority people to Han ways of life, but on the other, millions more resist transculturation and seek to sustain Buddhist, Muslim, or other ways of life.

Certainly the numbers are in favor of the Chinese empire. Few minorities are as clearly defined culturally as the Uyghurs or the Tibetans, and in many parts of southern China, where most of the non-Han ethnolinguistic groups are located, the cultural border between minority and Sinicized is vague and mobile. By some estimates, approximately 8.5 percent of China's population may be categorized as minority—less than half of Russia's more than 19 percent. But keep in mind China's human numbers: 8.5 percent of 1.3 billion people amounts to over 110 million. In actual terms, that is almost four times as many minority people as Russia rules in its vast territory. Like their Soviet counterparts, China's communist rulers inherited an empire from their predecessors, and the Chinese imperial mindset survived the transition.

LIVING WITH GOLIATH

Superpowers have imperial ambitions. Historians, historical geographers, political scientists, and other scholars have analyzed the ways these ambitions have changed from the era of colonialism to the period of globalization: the days of colonial acquisition and control are over, but not the urge to dominate in other ways. In the eyes of billions in the global periphery, globalization is shorthand for economic, political, and cultural intrusion propelled by powers in the core. From exclusive financial districts in central cities to gated communities in the suburbs, its manifestations are seen as a new version of old imperatives. It has nothing to do with ethnicity. It has everything to do with power and privilege.

It is certainly the case that European colonialism and Western neo-colonialism transformed the world, but these were not the only examples of the phenomenon. In the July 25, 2004 edition of the *New York Times Book Review* a group of eminent reviewers, discussing imperialism, proclaimed this to be an essentially Western phenomenon (apparently not only China but also Japan was off their cartographic radar); John Lewis Gaddis referred to the United States as "the only empire left." Obviously none of these experts was a geographer, but the exchange of views made one shudder. None of them was on record anticipating the fall of that other "Western" empire 15 years earlier, the Soviet Union; all seemed to agree that China had no imperial dimensions or ambitions.

As Ross Terrill points out in his book *The New Chinese Empire* (Terrill, 2003), not only is China the product of empire, but as a nation it is acutely aware of the losses it has sustained in what it still regards as its sphere of influence beyond its current boundaries. When China's last imperial dynasty was severely weakened, foreign powers forced border treaties on it that China's present leaders disavow; for example, Chinese maps show the northern part of the Indian State of Arunachal Pradesh as Chinese territory. In recent years a small area immediately to the east of Bhutan called Tawang has become an objective for Chinese expansionism. The justification for this is telling: because the people there are Buddhists, and China rules not only Tibet (Xizang) but also Buddhist communities in Qinghai, China should also control this particular pocket of Buddhism, even though it is clearly within India. From the "stolen lands" now under Russian control and the unsettled border with Kashmir to islands in and waters of the East and South China Seas, China has unfinished business in its Qing-era domain (see Figure 8-2). Mongolians cast a wary eye southward when they hear of Mongol-Han discord in the Nei-Mongol Autonomous Region.

None of this is surprising or alarming. The Chinese themselves often say that they have no imperial ambitions although they tend not to acknowledge the opposition their misrule in Tibet, Xinjiang, and Nei Mongol arouses; their negotiated agreement in 2011 with Vietnam over island claims in the South China Sea evinces a desire to avoid confrontation. In late 2011, when Myanmar's "civilian" generals seemed to want to steer a more independent course, China appeared to be willing to accommodate this—after having made huge investments in what was long seen as a virtual colony of Beijing. Wary the Mongolians may be, but China has made no overt claims to its Qing-era dependency. If current policies continue and China's domestic economic and demographic challenges grow, it is unlikely that territorial expansionism will become a larger issue.

While many countries have territorial issues with neighbors, these tend to take on greater significance when one claimant is already a giant. For many years the United States has been the long-term stabilizing force in the western Pacific, from Japan to Australia and from the Philippines to Thailand. America's postwar relationship with Japan, fostering democracy there, created the setting for one of the twentieth century's greatest economic success stories. The American military presence protected South Korea, prevented an explosion over Taiwan, calmed tempers over small islands and helped constrain North Korea's nuclear ambitions. America's military presence in the Philippines until 1991, ending when Mount Pinatubo's giant eruption destroyed air and sea bases on Luzon Island just as the Philippine government was debating the continuation of the policy, played a role in mitigating maritime disputes in the South China Sea. And Washington's close relationship with Singapore has been another element in this geopolitical framework.

But times change, and change seems to be happening ever faster. Even as China's perimeter remains comparatively calm and American leaders are discussing options for the relocation and partial removal of American forces from Japan and South Korea (President Obama in 2012 announced plans to remove 10,000 U.S. troops from Okinawa), other issues intrude. While the United States remains the world's unmatched military superpower, China's military is rapidly modernizing and expanding. High-level military meetings have exposed some strong differences of perspective and opinion between Chinese and American representatives. China casts a wary eye on Japan's bolstering of their antimissile capacity in the face of North Korea's nuclear program and missile tests; Chinese leaders frequently complain about the asymmetry between the U.S. presence in East Asia and the Chinese absence from the American side of the Pacific. But this, too, is changing. And managing this change, to avoid at all costs a rise in discord between America and

China that might spin out of control, should be the highest-priority foreign-policy objective on both sides of the Pacific.

MEETING CHINA IN THE WIDER WORLD

Visit China, even after a relatively short absence, and you will be amazed at the pace and scale of its transformation, often described as the greatest public-works project in the history of humanity. Bridges, tunnels, super-highways, airports, power plants, fast-train tracks, skyscrapers, factories, shopping malls, port facilities, dams—it is all bewildering and some-times overwhelming. Certainly the Chinese believe in their ability to overcome all obstacles through technology and megaprojects. The Three Gorges Dam is only one example of this, one of those "biggest in the world" achievements the Chinese routinely refer to now (it used to be an American habit, as we all remember). In 2012 the Chinese were plan-ning another almost unimaginable project: the diversion of one of the headwaters of one of their four major river systems into the catchment of another. The Huang (Yellow) River, which reaches the sea not far from Beijing, is losing water while the Yangtze, the river of Shanghai, has plenty. So why not change the course of the upper Yangtze (called the Chang in its headwater area) and feed it into the Huang?

To physical and environmental geographers this would be so fantas-tic, unbelievable, even unimaginable a plan, with so many unknowns and uncertainties, that it would hardly be taken seriously. But in China, no megaproject seems to be unattainable. This is not to say that there won't be opposition—some scientists argued against the Three Gorges Dam project—but if the national planners want it, it is likely to happen. Stay tuned; in a time of climate change, this could be the ultimate gamble.

The lesson is this: the Chinese have transformed their Pacific Rim, have given new meaning to the concept of a Great Leap Forward (and without the casualties of Maoism), and now their oyster is—the world. No longer are the Chinese absent from the American side of the Pacific: they are contracted to operate and modernize ports at both ends of the Panama Canal, they are building facilities at the Peruvian port of Callao to handle commodities that will someday travel through tunnels and across bridges from Brazil, they are Chile's main trading partner, they are buying the loyalties of Caribbean countries that used to support Tai-wan, they are forging ties with Venezuela and are buying Canadian oil, offering to build terminals on the coast of British Columbia. Even as China leaps ahead, the Chinese have fanned out across the globe, buying commodities and mines, building railways and roads, purchasing huge swaths of land, laying pipelines and forging strategic links.

If this sounds familiar to anyone who has studied the economic geography of colonialism, many observers are indeed comparing China's global activities to those of European colonial powers: in Africa, the Chinese are sometimes referred to as the "new colonizers." Like the colonial powers, China is buying commodities and bringing them home; for example, China is Angola's largest trading partner in large measure because of the oil it buys in that former Portuguese dependency. China is a major player in the economy of Zambia, where copper is the crucial commodity. China is engaged with Tanzania where it is repairing the (originally Chinese-built) TanZam Railway. China is Sudan's main customer (oil again), figures prominently in both import and export data for South Africa, and takes half of the exports of the DRC (Congo). And so it is in long-independent states. China is Australia's leading trading partner—indeed, China drives Australia's current boom economy. China also has become Brazil's biggest customer as well as supplier.

And it is not just oil and commodities. With ample funds at their disposal, Chinese investors are buying up land, securing long leases, and signing long-term production agreements with farmers that will lock the producers into lengthy commitments. So aggressive are the Chinese that several countries have begun to make it more difficult for foreigners to acquire land outright or secure long leases. In the Philippines, the Chinese sought a long lease to 1.2 million hectares (3 million acres) of good farmland, a "land grab" that got worldwide attention. Argentina, Brazil and a number of other countries have begun to limit purchases while still allowing leasing. In Argentina, a Chinese group in 2010 tried to buy a tract of 200,000 hectares (500,000 acres) of farmland before they agreed to lease it. But from Madagascar to Laos and from Kazakhstan to Colombia, millions of hectares of farmland are coming under Chinese control.

The consequences are mixed. Some economic geographers describe Brazil's relationship with China as a clear case of neocolonialism: in 2010, more than 80 percent of its exports to China were raw materials, while nearly all of Brazil's imports from China were manufactured goods, with a huge negative impact on Brazil's own industries. On the other hand, the growing trade in soybeans has lifted hundreds of thousands of farmers out of poverty.

Relations between Chinese visitors and settlers on the one hand, and locals on the other, also display a tinge of neocolonialism. Chinese entrepreneurs control the terms of trade; locals do the work, although Chinese firms also send large teams of Chinese workers when conditions warrant it. Several hundred Chinese workers, for example, are working one of the world's largest copper deposits in Afghanistan, protected by American security. When Chinese workers were brought in to work in Zambia's copper mines, serious friction resulted in violence and sabotage. In truth, Chinese workers and investors tend not to mix with locals the way European colonialists did—in Africa, South America, or anywhere else. The Chinese are often seen as aloof, preferring to stay within their own compounds, whether in Vietnam, Myanmar, Tanzania, or Brazil.

Wherever you look—at sea, on land, in the cities and in the countryside, in the mines and on the farms, China's global presence is evident. New Zealand's sheep-raising industry is dwindling under the Chinese quest for pastures on which to feed cattle to produce milk—converted into the milk powder that is exported in volume to China's market. Iceland made the news in mid-2011 when a Chinese entrepreneur sought to buy .3 percent of the country, seemingly a tiny sliver but, if it were the United States, a piece of land the size of Massachusetts.

Several scholars read the tea leaves of Chinese assiduousness and interpret it as a sign of things to come: world domination. In his book *When China Rules the World*, Martin Jacques argues that China, as the world's leading "civilization state" will, in the language of its subtitle, cause "the end of the Western world and the birth of a new global order" (Jacques, 2009). The question is only one of time and method: assuming that war is avoided, China's inexorable rise will constrain and hamstring a United States (and West) that have been used to freedom of action of the kind last seen in Iraq and Afghanistan. None of this, of course, adds up to a cold war—yet. But the perceived twenty-first-century American penchant for unilateralism, impulse, and proselytism is creating conditions that could contribute to its emergence. During the Cold War of the twentieth century, each side was convinced of the other's nefarious designs, and each side saw the other as morally wrong: communist perceptions of the inequities of capitalism were countered by capitalist notions of atheism and repression in the "evil empire." So it comes as an unpleasant surprise to many Americans that the one-time guarantor of global stability is now viewed by many as the more likely protagonist in a Transpacific Cold War.

Are the United States and China on such a collision course, and if so, what are the respective causative and mitigating factors? Let us look again at the political geography of China. On the perilous side, there are China's expansive past and imperial present, its communist-authoritarian governance, its dreadful human rights record, its demographics (the one-child-only policy is creating a surplus of tens of millions of males), its world's-largest military, its fast-rising nationalism, its growing demand for global raw materials including oil, its unsettled relations with Japan, its designs on Taiwan, and its problematic role with regard to North Korea, its communist neighbor with a terrorist history and nuclear ambitions. On the mitigating side, China, unlike the Soviet Union, does not overtly seek to export its communist system or ideology except to SARs and Taiwan, maintains a strictly secular society, shares with the United States a concern over Islamic terrorism, has opened its doors to economic development, has settled some territorial issues with neighbors, and has withdrawn far-reaching maritime claims.

China is ascending in a world dominated by a superpower that, during the twentieth century, combined unmatched altruism toward former enemies with still-inexplicable Cold War foreign-policy decisions. Even as the United States brought democracy to (West) Germany and Japan and massive financial assistance to Europe, held the nuclear-armed Soviet Union at bay and opened the door to China's reentry into the world, Washington, like Moscow, supported repressive regimes from Middle America to Monsoon Asia that made Myanmar's generals seem like model democrats. The end of the Cold War led to the end of many of those dictatorships, but the new threat of Islamic terrorism, in part the direct result of an earlier foreign-policy error in Afghanistan, created a new set of circumstances only the outlines of which are visible today. Although there is much concern about the supposedly novel unilateralism in American policy today, this is nothing new. The United States may not have signed on to the Kyoto protocols or other recent treaties, but neither did it ratify the much older UN Convention on the Law of the Sea (UNCLOS) or a number of other international agreements with pre-2000 datelines. What is new, however, is worldwide American interventionism in response to the 2001 terrorist attacks in New York and Washington, a byproduct of which is an inconsistent campaign to install or advocate democracy in countries ruled by authoritarian regimes (inconsistent because it exempts stronger regimes such as China as well as others whose resources are crucial to the American economy, such as

Saudi Arabia). In this context the 2003 intervention in Iraq, with its twin aims to depose a ruthless dictator and to establish a democratic government, had far-reaching consequences. To the generals in Myanmar and the despots in North Korea, China's geopolitical ascent in the Asian perimeter was a welcome and timely development. To other countries in the region whose human-rights records or democratic institutions might not meet American criteria, China's deepening economic involvement provides a counterweight to United States pressure.

Is all this enough to precipitate a cold war? Not yet, but if Chinese public opinion is reflected by what is heard in college classrooms, read in the press, and seen on the Internet, the prospect has risen in the nation's consciousness. American criticisms of China's human-rights deficiencies are met with angry denunciations of the failures of American race relations; American commentaries on China's undemocratic system of government are countered with critical analyses of the shortcomings of democracy in the United States. Public support for United States intervention in Afghanistan turned into widespread opposition to the invasion of Iraq, and in my experience disapproval of the involvement of South Koreans in that campaign was nearly universal in China. In the process, politics has begun to displace economics as the chief topic of discussion and debate, which is not a good sign for the future.

The most populous nation on the planet, heirs to a great empire and guardians of one of the oldest continuous cultures, is asserting its place in a world dominated by American superpower. China is the first non-Western power to mount such a challenge, creating a new set of geopolitical circumstances. During the twentieth century, when the United States and the Soviet Union were locked in a Cold War that repeatedly risked nuclear conflict, Armageddon never happened because this was a struggle between superpowers whose leaderships, ideologically opposed as they were, understood each other relatively well. While the politicians and military strategists were plotting, the cultural doors never closed: American audiences listened to the music of Prokofiev and Shostakovich, watched Russian ballet, and read Tolstoy and Pasternak even as the Soviets cheered Van Cliburn, read Hemingway, and lionized American political dissidents. In short, this was an intracultural Cold War, which reduced the threat of calamity. A cold war between China and the United States would involve far less common ground, the first intercultural cold war in which the risk of fatal misunderstanding is incalculably greater than it was during the last.

How can such a cold war be averted? Trade, scientific, cultural, and educational links and exchanges are obvious remedies: the stronger our interconnections, the less likely deepening conflict becomes. We Americans should learn as much about China as we can, to appreciate its experiences and to understand the historic and cultural geographies underlying China's views of America and the West. China today is a major player on the world stage, internationally involved and globally engaged. Constraining the forces leading to a world-power competition between the United States and China will be the key geostrategic challenge of the twenty-first century.

9

EUROPE: SUPERPOWER IN THE MAKING OR PAPER TIGER OF THE FUTURE?

MANY AMERICANS THESE DAYS REGARD EUROPE WITH A CERTAIN ambivalence. As the "Greatest Generation" that fought alongside British, French, and other European allies in World War II fades into history, a younger constituency casts a wary eye on a Europe that doesn't seem to be what it was. The French, in particular, proved a disappointment before and during the Iraq War, when they would not support the Bush-Cheney intervention, and some Americans took to ordering and eating "freedom fries." The Germans had previously failed in their role during the collapse of Yugoslavia (but German beer remained popular). A United States Secretary of Defense publicly scolded the "old" Europe for its shortcomings while forecasting a bright future for the "new"—that is, the east, far from the Italians and the French and the Dutch and the Germans. American military leaders expressed regret over the small size of European contingents helping in the campaign in Afghanistan, and the restrictions placed by European governments on the tasks they would be allowed to perform. In 2011 another frustrated Secretary of Defense warned that if European members would not show stronger financial and logistical support for NATO, the alliance risked sliding into irrelevance. Meanwhile conservative American voters saw a "socialist" Europe even as liberals criticized the growing strength of right-wing parties from Britain to Bulgaria. And when Greece came to symbolize ineffective government and a bloated public-service sector, its budget a shambles, its streets in chaos, and its debts a Transatlantic risk, American doubts about their erstwhile ally only grew larger.

In truth, many Europeans are ambivalent about Europe as well. This has to do with other things: for nearly three generations, Europeans have been in the grip of a combination of social, political, and economic forces that have both united and divided them. Ever-greater unification, Europeans will tell you, is the dream of the elites, of the bureaucrats, of governments that know better than "the people." Ever-more rancorous division separates these dreamers from the workers, who hate to see their taxes go to an experiment whose value and objectives they doubt. The British, many of whom do not see themselves as Europeans, may be the most Euroskeptic of all, but every country in Europe has its naysayers. In the process, it has become difficult to geographically define just what Europe is: the group of countries united under the umbrella of unification? All the countries lying west of Russia? The countries of Europe *including* Russia?

AT THE HEART OF THE WORLD

For centuries, Europe lay at the heart of the human world. European empires spanned the globe and transformed societies far and near, for good or ill. European capitals were the nodes of trade networks that controlled distant resources. Millions of Europeans migrated from their homelands to the New World as well as the Old, creating new societies from America to Australia. Long before globalization became equated with Americanization, it was a process of Europeanization.

In agriculture, industry, and politics, Europe generated revolutions—and then exported those revolutions throughout the world, consolidating the European advantage. Yet during the twentieth century, Europe twice plunged the world into war. In the aftermath of the Second World War (1939–1945), Europe's weakened powers lost the colonial possessions that for so long had provided wealth and influence. European countries were threatened by communist parties and movements, and an ideological Iron Curtain from the Baltic to the Adriatic split the continent apart. East of it, Soviet communism dominated from its headquarters in Moscow. To the west, liberal democracy and market capitalism prevailed, but with crucial help from Washington and not before a few dreadful dictators left the scene (Hitchcock, 2002).

Western Europe's economic recovery and Eastern Europe's eventual rejection of communism were this realm's dominant events over the last half century, but another story continues to make the headlines. Europe's countries are engaged in an unprecedented experiment in supranational-ism, a process of international unification and coordination in numerous spheres ranging from the economic to the political. It is an experiment that some leaders hope will ultimately lead to the formation of a United States of Europe, a federal superpower that will constitute a counter-weight to the global dominance of the United States of America.

Ordinary Europeans, perhaps a majority of them, do not always share this enthusiasm for unification. But the process has gone further than could have been imagined 60 years ago. Europeans can drive from Lisbon to Lithuania without showing their passports at national borders. They can use the same currency, the euro, in 17 countries (as of late 2011). And the European Union (EU) continues to expand. In 2004, it experienced its greatest growth surge as 10 countries joined 15 members to create a 25-member organization; then in 2007 the EU reached the Black Sea when it admitted Romania and Bulgaria. And the process is by no means finished. Croatia joined in 2012, and other states emerging from the collapse of Yugoslavia were candidates as well. For all its foibles, the EU continued to attract would-be members.

ASSETS AND LIABILITIES

Will Europe achieve the supranational and superpower ambitions of its leaders? Look at the map, and Europe appears to be a mere protrusion of peninsulas extending westward from the great landmass of Eurasia. Read the scale, and you see how small Europe is territorially, not much larger than the United States west of the Mississippi (Texas is almost twice as large as Germany; the United Kingdom is about the size of Oregon). And Europe's territory is anything but compact. Peninsulas and islands large (by European standards) and small are flanked by seas wide and narrow. Europeans gazed at each other across straits and channels, bays and rivers from the day they settled here.

That settlement came in waves from the east as intermittent warming during the Wisconsinan Glaciation enabled the first modern humans to survive the rigors of Europe's variable climates. This invasion started slowly, but *Homo sapiens* could do what their Neanderthal predecessors could not: mix their hunter-gatherer existence with the pursuit of migrating reindeer, and adapt when the climate turned colder. Modern humans have been in Europe for more than 40,000 years, moving into Iberia and Italy during the coldest times and spreading northward when

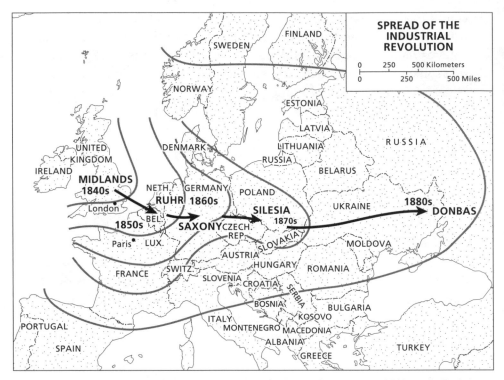

Fig. 9-1. The Industrial Revolution spread eastward from the British Midlands in the 1840s to the Donbas Region on the border of Ukraine and Russia nearly a half century later. It transformed Europe's—and the world's—economic geography.

it got warmer. But the real population expansion came during the current interglacial warm period, starting about 12,000 years ago and interrupted only by the brief Younger Dryas cold spell. Wave after wave of immigrants moved from Eurasia's interiors toward the warmth of the west and the isolated security of peninsulas and islands along its coasts. They brought with them their diverse cultures and varied languages, of which the earliest are remnants now along Britain's and Ireland's western shores. To this day some of the routes taken by these early arrivals are uncertain; no one knows, for example, how the Basques reached their present homeland on the Spanish-French border because their language has no links to any other. The Hungarians, Estonians, and Finns also arrived from yet-unknown sources and planted their languages in Europe. At least their familial relationships are clearer—although a recent study suggests a link between these Uralic languages and . . . Japanese.

The dominant languages of Europe today, including English and German, Spanish and French, Polish and Ukrainian, belong to the several branches of the Indo-European language family, and from their present distribution you can infer their genesis. The speakers of the Germanic languages, also including Dutch and the Scandinavian tongues, spread westward across the North European Lowland, the vast plain that extends from the shores of the North Sea into Russia between the Alps to the south and the Baltic Sea to the north. They invaded Britain and Ireland and displaced the older Celtic-speakers there; as Europe warmed, they settled in Scandinavia and even Iceland. The Romance languages, including modern Italian and Romanian, evolved from the Latin spoken by the Tiber Valley peoples who founded and forged the Mediterranean-encircling Roman Empire and who welded their language to the tongues of its provinces. And the Slavic languages, including Czech and Bulgarian, are the languages of the east, latecomers in the Roman-Mediterranean sphere but long established in neighboring Russia.

Superimpose Europe's language map on its map of physical landscapes, and you see one major reason why this realm remains a Tower of Babel. The gentle relief of the North European Lowland (which extends into Britain as well) facilitates movement and interaction, and here lie the three historic powers and modern economic powerhouses of the realm: the United Kingdom, France, and Germany. But even here, real life is more complicated than the map suggests: major languages are not mutually understandable and minor tongues (such as Frysian) and strong dialects inhibit comprehension and reciprocity. Where the relief is higher, notably where the Alpine System prevails, physical barriers to interaction contribute to isolation and pervasive cultural fragmentation. Before its disastrous disintegration, the former Yugoslavia, about the size of Wyoming, had 24 million inhabitants who spoke 7 major and 17 minor languages. Peoples living in adjacent valleys often had no effective contact with each other, separated as they were by mountain spurs, language, and tradition.

Seen in this context, it is amazing that European integration has proceeded as far as it has. But Europe's cultural diversity is more than matched by its natural diversity, and whereas the former is a challenge, the latter presents opportunity. From the flat coastlands of the North Sea to the grandeur of the Alps, and from the moist woodlands and moors of the Atlantic fringe to the semiarid prairies north of the Black Sea, Europe contains an almost infinite range of natural environments. The insular and peninsular west contrasts strongly with the more continental, interior east. Dry-summer Mediterranean climate in the south yields to year-round moisture on the North European Plain and cold-winter regimes in Scandinavia. Crops change from oranges and olives to fruits and vegetables to grains and potatoes. Atlantic warmth and moisture lose their effect into the continental interior, where the crop pattern changes again. Europeans have been trading for millennia; the Romans made their Mediterranean Sea an avenue of commerce.

There is more. Europe may be small, but the range of its mineral and energy resources is large. A backbone laden with raw materials extends across middle Europe from the coalfields of England to the iron ores of Silesia, propelling the Industrial Revolution when the time came (Fig. 9-1). And, as the ancient Romans already knew, there are pockets of valuable minerals ranging from copper to gold in the highlands and mountains from Spain to Scandinavia. For centuries, individual places on the European map were known for their specialized products, often based on such locally available resources. We still carry our idiomatic coal to Newcastle and advertise Italian marble in luxury homes; Sheffield was long known for steel the way Detroit was for automobiles. From resources below and atop the ground Europeans made products that were peerless on world markets: Swiss watches, Dutch cheese, Irish linen, French wines, Swedish furniture, Finnish electronics. Europe's domestic heavy industries produced trains and ships, cars and planes, trucks and tanks.

The European stage on which all this happened may be tiny, but it is very crowded—even after the departure of millions of emigrants headed for the New World. Americans tend to be surprised when they discover that Europe's total population is about double that of the United States, nearly 600 million in 2012. Crowded Europe also is one of the world's most highly urbanized realms, especially in several countries of the west where 90 percent or more of the population lives in cities and towns. Europe's great cities, from London to Rome and Paris to Athens, carry the imprints of Europe's turbulent past and tumultuous present in their historic centers, clustered, space-conserving neighborhoods, and immigrant-laden outskirts. They do not display the profligate suburbanization of American cities, so that sprawl is comparatively limited and public transport more effective. But internal circulation is hampered by old, inefficient street patterns and narrow roadways. As a result, highways linking major urban centers are the scenes of some of the world's longest traffic jams. Where the modern highway meets the clogged arteries of the historic city, the slow dispersal of arriving vehicles can back up traffic for dozens of kilometers.

Europe's population is not only highly urbanized; it is also old. The populations of nearly half of Europe's countries are shrinking, and those of most others are growing very slowly. People living in crowded cities tend to have fewer children than their rural counterparts, but other factors also contribute to Europe's population stagnation. Later marriage, high unemployment, the cost of child rearing and other financial uncertainties, and the breakdown of religious strictures on family planning all play their role (Italy, of all countries, has zero population growth today). The implications of this for Europe's future are troubling: younger taxpayers must pay for the social services required by older citizens, and the shriveling of that taxpayer base confronts European governments with some difficult options. Combine this with the continuing influx of immigrants from all over the world, and you can see why Europe's prospects are mixed.

WHERE AND WHAT IS EUROPE?

Silly question, it might seem. Europe, of course, is the British Isles and Scandinavia and Greece and Poland . . . but maybe not Moldova or Belarus?

Welcome to a long and probably endless geographic debate. The location of the eastern boundary of Europe has been the subject of argument for many years. The key issue has always been whether Russia is part of Europe or not. When the Soviets made satellites out of much of Eastern Europe, Russia was seen by many Europeans as an external colonizer, especially when communist and noncommunist Europe went vastly different ways politically as well as economically. When the Soviet empire collapsed and Russia emerged as a fledgling democracy, there were visions of a Europe from Madrid to Moscow and beyond.

But that's just the problem. If Russia is a European state, does Europe therefore extend from the Atlantic to the Pacific, from London to Vladivostok? Not according to the cartographers who worked on the eighth edition of the *National Geographic Atlas of the World* (National Geographic Society, 2005). On page 71 their map of Asia shows a green line (a "commonly accepted" boundary, a note states) running along the Ural Mountains, then cutting across western Kazakhstan, across the Caspian Sea to turn westward along the Caucasus Mountains, and then across the Black Sea to the Bosporus. Europe's eastern border, according to this convoluted construction, largely coincides with the Ural Mountains, and there is a European Russia and a non-European Russia. It must seem rather strange for the people of the city of Ufa, just west of the Urals, to know that they live in Europe—while their compatriots in Chelyabinsk, just down the road, live in Asia.

The National Geographic *Atlas* has a lot of geographic clout, and millions of readers must have taken this notion seriously. But this version of Europe's dimensions makes no sense. If Russia is part of Europe, then Europe extends all the way to the Pacific, from Scotland to Sakhalin. If it is not, then Europe's eastern border coincides with Russia's western one.

There are many reasons to adopt this last solution. Make a list of Russia's geographic properties, and the differences between it and its European neighbors leap from the page. Russia is 100 times as large territorially as the average European country. Russia's population is nearly twice as large as Europe's largest. Russia's mainly energy-export economy is unlike that of Europe (you are unlikely to find Russian specialty products among your purchases). Russian democracy is still rickety, and two neighbors over which Moscow has residual influence, Ukraine and Belarus, retain authoritarian habits, although Ukraine is showing signs of reorientation. Russia is not even being mentioned as a potential member of the European Union. Certainly Russia's cities and cultural landscapes have European overtones, but that is not enough to conceive of Russia as a functional part of a European geographic realm.

Since we're talking about regional boundaries, what about Europe's southern border? It seems logical to regard the Mediterranean Sea as Europe's southern limit, but that's only because of what happened after Roman times. Some of the westward migrants who arrived in this part of the world took a southerly route around the Mediterranean and ended up in what is today North Africa. There they were overpowered (or ousted) by the Romans, whose provincial administrations transformed the area and integrated it into the European orbit. Rome's collapse might have merely delayed North Africa's Europeanization except for one crucial event: the arrival of Islam and the reorientation of North Africa to Mecca. By the time the French, Spanish, and Italians arrived to colonize North Africa, the Mediterranean—once Rome's Mare Internum—represented an unbridgeable cultural chasm.

Does all this matter? Absolutely. To be part of Europe means that a country can hope for access to Europe's many international economic and financial organizations, for representation in Brussels and Strasbourg, for mutual security, and for many other advantages resulting from cooperation among neighbors. The name Europe today stands for far more than a continent or a geographic realm. It also represents international opportunity and progress. Make no mistake: Europe has its much-publicized problems, its ambitious unification program creaking under economic difficulties and political discord. But collectively, the 50 countries of Europe represent a socially advanced, economically prodigious, politically mature cluster of stable states whose combined productivity ranks in the world's top three, whose governments (with only Belarus excepted) are representative, and whose living standards rank among the highest in the world. Few countries would forgo the chance to join the European Union (the Icelanders, Norwegians, and Swiss are in a small minority, and each is having second thoughts). True, the current (2012) global economic crisis may produce setbacks, but the great EU experiment will continue: the list of hopeful applicants keeps growing (Fig. 9-2). After Croatia, joining in 2012 as the 28th member, look for Serbia and its neighbors, Ukraine, and even some countries not geographically part of Europe—Georgia and Turkey—to aspire to membership.

In the process of its tumultuous postwar reconstruction, Europe has been redefined. Today there is not only an EU currency but also an EU passport. When travelers arrive at an EU airport, they are directed *not* to a British or French or German immigration booth, but to an EU (or non-EU) waiting line. In every member state, the EU flag, twelve yellow stars on a dark-blue background, flies alongside the national flag. Nationalists in all member countries may be irritated, but Europe has achieved a level of multinational integration unimagined when the experiment began. How it got there is the story of this chapter.

FRACTIOUS EUROPE

In 2003 I was asked to make a presentation about Europe as part of the ceremonies marking the launch of a cruise ship in Southampton, England. The invited audience of several hundred was international, roughly consisting of an equal number of Europeans and Americans. I spoke about Germany's then-serious economic problems, France's quarrel with the United States over Iraq, prospects for the euro and EU enlargement, and

Fig. 9-2. The European Union: the Ins and the Outs, the past and the future.

the issue of a European Constitution, then very much in the news while it was being prepared, a momentous event in the EU's history. I went on too long and left no time for a Q&A session, but asked anyone with comments to come up to the lectern afterward. Soon a group of about a dozen listeners converged on me, and I could see that some of them were quite angry. "You were unfair to Germany's government!" shouted a man in the middle of the pack. Before I could answer, someone started a bitter complaint about my view of the French. "No," said the vociferous German, "he was quite right about you French. You want to run the European Union, but the British won't let you do it." In a few moments the Europeans among the group were in a shouting match with each other, no longer interested in arguing with me. When I quietly left the room the dispute continued undiminished.

Looking at the fractured political map of Europe these days that episode comes to mind. If you don't count Europe's microstates (such as Monaco, Andorra, San Marino) or nonstate territories like Gibraltar and Greenland, there still are over 40 countries in Europe, a jumble of states creating as complex a political mosaic as any in the world. In some parts of the realm, for example along the Dutch-Belgian border, small parcels of land belonging to one country are completely surrounded by territory of the other. Europe's political map is a legacy of centuries of conflict and adjustment.

And the conflicts are not over. Basque extremists in Spain until recently killed members of government, judges, and policemen to stake their claim for independence. Several hundred thousand people died during the 1990s as Christians and Muslims, Serbs and Croats destroyed much of Yugoslavia's historic heritage. Until just a few years ago Catholics and Protestants in Northern Ireland fought and died as though the sixteenth century never ended. Corsicans used terror to promote their cause against the French. Other movements, many of them regional in nature, have the potential to engender strife. In February 2005, Montenegro launched its successful bid to negotiate secession from Serbia. Kosovo's split from Serbia in 2008 entailed agitation and some violence. Tensions over political status, cultural issues, or historic injustices (often in some combination) still afflict many European countries (Fig. 9-1).

Europe has long been a crucible of culture, but it is also a cauldron of conflict. Twice in the twentieth century, Europe plunged the world into war, leading to the use of weapons of mass destruction both times (gas in the first, atomic bombs in the second). Twice the combatants came out of the conflict saying "never again." But when Yugoslavia disintegrated in the 1990s, and Europeans had the opportunity to prove that they meant it, they failed. In Europe, never say never.

Against this background, Europe's success in overcoming its historic barriers and enmities is encouraging—not just for Europeans, but for states and nations all over the world. For years now, Europeans have been working to lower obstacles to cooperation and to facilitate the free flow of people, products, and money. This has become the world's greatest experiment in supranationalism, in which states voluntarily yield some of their sovereignty in the interest of the common good, and lessons learned in Europe have been eagerly applied elsewhere. Today there are about 40 sometimes-overlapping supranational organizations among the approximately 200 countries of the world, some of them quite consequential, such as NAFTA; others, such as Russia's almost-forgotten CIS and Africa's "African Union," of little impact as yet. For Americans, the lesson of NAFTA is that supranational cooperation has its costs as well as profits. American jobs lost to Mexico became a hot political campaign issue as soon as NAFTA got started. And NAFTA is no EU. Europeans have given up a lot more than job security to make their union work.

THE TRUMAN PLAN

Lest we forget (some Europeans don't always remember), it was American generosity that got the European unification movement off the ground. President Harry Truman's secretary of state, George C. Marshall, in a speech at Harvard University on June 5, 1947, proposed a huge United States investment in Europe's recovery. It was not all altruism, to be sure: the American administration feared that communist parties would gain control over European countries west of the Iron Curtain, and helping Europe recover economically would pay huge dividends politically as well. Under the terms of the Marshall Plan (which might well be called the Truman Plan), European countries—even those behind the Iron Curtain—were asked to propose a massive self-help program that would be funded by the United States. Sixteen European countries, later joined by defeated (West) Germany, presented a plan under the title of Organization for European Economic Cooperation (OEEC). The Soviets forbade any of their Eastern European satellites from participating, so when the United States Congress funded the OEEC to the tune of $12 billion (about $140 billion in present dollars), all of it went to the West.

European enthusiasm for the Marshall Plan grew from the realization that it would have political as well as economic consequences. By enmeshing all major European states in this multinational scheme, the risk of a third war would be minimized, and Europe could set about its recovery under the security of the military North Atlantic Treaty Organization (NATO). The Marshall Plan commenced in 1948 and the protective shield of NATO took effect in 1949. Now it was up to the Europeans to convert their good fortune into lasting cooperation and to accept a rehabilitating (West) Germany as part of the mission.

The Marshall Plan lasted for four years (1948–1952) and set the stage for developments probably unforeseen even by the most optimistic European leaders (Rifkin, 2004). The Council of Europe, just a deliberative body but viewed as the forerunner to a European Parliament, was created in 1949. A crucial economic step was the formation in 1951 of the European Coal and Steel Community, removing trade barriers and establishing a common market for coal and steel among the six member states. When the Marshall Plan phased out in 1952, the momentum toward European integration never slowed. By 1957 six of the seventeen aid recipients were ready to ratify the Treaty of Rome that, in the following year, launched the European Economic Community (EEC), the so-called Common Market or "Inner Six."

SIX TO NINE TO TWELVE

Why only six? It was the old story of European divisiveness. While France, Italy, (West) Germany, and the three Benelux countries were ready to take the next step, the British felt that their future was more closely linked to the Commonwealth and they did not want to risk those ties by joining the EEC. On the other hand the British did want to keep a stake in Europe, and so they founded an organization to parallel the EEC consisting of the United Kingdom, the three Scandinavian countries, the two mountain states (Austria and Switzerland, the latter an unlikely joiner), and Portugal. This group was known as the European Free Trade Association or "Outer Seven," but it was no match for the powerful six of the Common Market. The leadership of the EEC was, to put it mildly, not pleased with this initiative. When the British changed their minds and applied for membership in the EEC, France vetoed their application. It was, so it appeared, European business as usual.

While all this wrangling was going on, certain European leaders wanted to remove the "economic" qualification from the EEC name. Europe should aspire to be more than an economic community of states; there would be other arenas of integration. And so, in 1967, the organization got its second name: European Community (EC). In 1968 the six EC members eliminated all internal customs duties and erected common external tariffs. This got the attention of the old EFTA group, and most of them, led by Britain, again applied for admission. This time the French held their fire, and in 1973 the United Kingdom, Ireland, and Denmark joined, while the Norwegian people rejected membership by referendum. The Inner Six had become "The Nine."

Behind the scenes, the fabric of the European Community got ever more intricate. The old OEEC was extended, and in effect superseded, by a broader, more international organization that now includes not only European countries but also the United States, Canada, Japan, Australia, and New Zealand: the Organization for Economic Cooperation and Development (OECD). Agreements were drawn up among EC member states ranging from agricultural policy to human rights and from monetary policy to labor regulations. One of the most important and portentuous of these agreements had to do with subsidies: the richer members were obliged to help the poorer ones by contributing to a fund established to reduce inequalities within the EC. On the political front, the year 1979 saw the first elections to the European Parliament that, in that year, had 410 members charged with legislative and consultative tasks. By 2012 that number had mushroomed to 754, amid much criticism of the EU's administrative cost, and before Croatia's delegation was seated.

And the European Community continued to grow. The admission of Greece (1981) and Spain and Portugal (1986) created "The Twelve"—a signal moment in the history of the unification movement in a number of ways. First came the ratification of the Single European Act, targeting a more integrated European Union (EU) in the 1990s—so now the organization was to acquire its third name in as many decades. Second was the acceleration of progress toward a single currency through a European Monetary Union (EMU). Third, plans were laid for the Maastricht conference of 1991 that was to lay the groundwork for further integration as well as expansion. Fourth, the need for an EU Constitution was recognized and addressed. And fifth, it was agreed that, no matter how many countries would join the EU later, the flag would carry twelve stars in the now-familiar circle.

Many observers and analysts doubted that European countries would be willing to dispose of their historic and emblematic currencies—the franc, the mark, the lira, the guilder—in favor of something to be called the euro. Surely the British would not abandon their pound? But the doubters were proven wrong: while the British kept their pound, the Germans, French, Italians, Dutch, and eight other countries in 2002 switched from their domestic currencies to the EU's euro. And while all this was going on, the EU, having already grown to 15 members, now

Fig. 9-3. Europe's historic fractiousness makes the European Union a remarkable achievement, but devolutionary pressures continue to simmer and sometimes boil. From Scottish independence (still contemplated as a goal by a significant minority) to Kosovo's turbulent secession from Serbia, European minorities push in one direction even as national governments move toward the other: still-stronger unification.

admitted ten more (see Figure 9-2), debated a hastily-written Constitution (rejected in 2005 by voters in France and the Netherlands), took in two more members in 2007 and negotiated with several other would-be members over eventual admission. If this sounds like haste making waste, well, the waste was not long in coming. By 2010 a financial crisis afflicted several EMU members, the EU was faced with the need for massive bailouts, and those who had advocated slower growth and tighter rules seemed to have it right.

TOO FAR TOO FAST?

You don't have to be a specialist in international affairs to realize that the expansion of a supranational organization across countries with ever-greater disparities in wealth and income, not to mention dimension and demographics, requires relaxation of the rules of membership. The old Common Market consisted of six fairly similar societies (well, five of them at least, and half of the sixth), and even the admission of the next three members kept the range of indices pretty narrow. But when Greece, Portugal, and Spain joined, the picture changed quite drastically. Not all the senior members of the EU were pleased, but soon funds were flowing from Brussels to Athens, Lisbon, and Madrid under the terms of the EU's program to help poorer countries and regions improve their infrastructures. Portugal built bridges and highways. Spain constructed a high-speed rail link between Madrid and Seville in poverty-stricken Andalusia. Greece erected a new airport near its capital.

The issue of depth versus breadth nevertheless roiled EU discussions. The longstanding objectives of the organization transcend economic integration and include common policies in foreign affairs and defense, and these seemed imperiled by enlargement. In practical terms, how could decision-making procedures originally devised for a supranational community of just six states, and modified only slightly since, function adequately for a union of 28 or more? How could the economies of poor countries such as the Baltic states be expected to meet the responsibilities and costs of EU membership after joining a fraternity organized for the wealthy? How many decades would it take for the new members to catch up with the old ones, given current economic growth rates? How much more difficult would it be for Europe to speak and act as one in the international arena when so many more voices—and votes—must be heard and heeded?

A key issue confronting the EU following its 2004 enlargement centered on the way expansion affected states' relative power. The ten countries joining in 2004 had some 75 million inhabitants, of which Poland had nearly 40 million and Malta only 0.4 million—a smaller population even than Luxembourg. The issue was how to assign votes in the European Council when the largest state has more than 200 times as many people as the smallest? The older, larger members were reluctant to yield some of their power and influence to the smaller, newer upstarts, but a method had to be found to satisfy all. An early plan, hammered out in

Nice, gave the two midsize countries, Spain and Poland, almost as many votes in the European Council (the EU's key governing body) as Germany, more than twice as populous. But during the drafting of the European Constitution in 2003, that plan was dropped in favor of one called the "double majority," which gave more votes to the more populous countries but required any legislation to be approved by (1) a majority of votes and (2) a majority of member countries, 13 or more until 2007 and over 14 after the pending admission of Romania and Bulgaria. When this proposal met further objection, a new plan defined a majority as consisting of at least 55 percent of member states representing 65 percent or more of the EU's population. This model would accommodate any expansion (or shrinkage) of the Union, and it satisfied all involved.

The same cannot be said for the European Constitution as a whole. Late in 2001, the then-15 EU leaders proclaimed a European Constitution desirable because, as the momentous 2004 expansion approached, the EU stood at a crossroads, a defining moment, in its history. To American observers, the EU's efforts to achieve consensus on something as far-reaching as a Constitution held much interest. The United States Constitution holds a special place in American life, and to witness historic allies struggle to achieve a consensus of this kind in modern times was fascinating. Quite apart from its practical side—the bureaucratic arrangements necessary to allow the EU to function through its expansion—the EU Constitution also, and more controversially, addressed sensitive issues ranging from the character of "European civilization" to Europe's "religious heritage." The draft text described the former as having "gradually developed the values underlying humanism: equality of persons, freedom, respect for reason," a construction unlikely to provoke much debate. But how to refer to Europe's religious heritage prompted much acrimonious discussion. In an earlier draft, the writers deleted specific references to Christianity as well as Greek and Roman civilization and even the Enlightenment in a compromise that sought to eliminate allusions to both religious and secular foundations of European civilization. Later, the word "religious" was inserted in a sentence that describes the "cultural, religious, and humanist inheritance" of Europe. That was not enough for representatives from Italy, Ireland, and Spain among older members and Poland, Lithuania, and the Czech Republic among the new ones, all of whom wanted some reference to Europe's Christian (or Judeo-Christian) heritage in the preamble. In truth, however, Europe, especially older Europe in EU terms, is increasingly secular; church

SAYING NO TO EUROPE . . .
BY LEAVING

Here's a trivia question: 28 states are members of the EU, quite a few are trying to get in, and some are thinking about it. But has anyone ever *left* this coveted club?

The answer is yes, though it wasn't exactly a state that gave the EU thumbs-down, and at the time the EU was still known as the EC. If you've never heard of Kalaallit Nunaat, that's probably because we tend to use the name Greenland for this Danish territory, once ruled jointly with Norway but a Danish dependency since 1814. Sensitive to indigenous wants and wishes, the Danes changed Greenland's status from colony to province in 1953, and then in 1979 gave it home rule under its distinctive Inuit name.

But in the meantime Denmark had joined the EC (in 1973), which meant that the inhabitants of "KN" had to abide by EC regulations, including fishing limits. The locals decided to test their home-rule prerogatives by announcing that they were withdrawing from the organization, and in 1985 they did.

This geographic story isn't over. It may be hard to believe, but Kalaallit Nunaat is moving toward independence, and if this happens, we'll see another demographic microstate (about 60,000 people) on the world map and in the United Nations. The prospect of oil reserves in adjacent waters and global warming shrinking the Greenland icecap is giving the Inuit notions of wealth and fortune—not to be yielded to a bunch of bureaucrats in Brussels.

attendance is declining and you see frequent references to "post-Christian Europe" in writings about the realm. Some framers of the European Constitution argued that the vigor of Islam in present-day Europe and Islam's historic invasions of Europe in Iberia and the Balkans justified the inclusion of "Judeo-Christian-Islamic heritage" as a defining feature of European culture, if there was going to be any reference to religion at all.

When voters in France and the Netherlands rejected the draft Constitution, the whole project seemed dead in the water; the British, among others, were reluctant to put the Constitution to a vote at all, because polls showed that here, too, the response would be negative. Still, those promoting the initiative have not given up, arguing that an EU Constitution, with its provisions for leadership positions and clear statements of common goals will strengthen the Union's position in the world.

EUROPEAN GOVERNANCE

The European Union today is a multinational entity with a population of over 500 million, exports that exceed 40 percent of the world's total, a common currency, a military umbrella defined by NATO, and a functioning (if excessively complex) governmental structure. It may not yet be a United States of Europe, but it is far more than a loose amalgam of culturally diverse states.

And it has a president and what Americans would refer to as a secretary of state. Given its global importance, you would imagine that these two European leaders would have names familiar to most of us. So have you heard of President Herman van Rompuy or "high representative for foreign affairs and security policy" Catherine Ashton?

Van Rompuy and Ashton were selected, not by popular vote but by the leaders of the (then) 27 member EU states, for positions established under the Lisbon Treaty, the revised framework for EU governance that went into effect on December 1, 2009. But you are not likely to see President Obama negotiating policy with President Van Rompuy; crucial negotiations involve leaders of key EU states, not the EU bureaucracy. Over time, the European Union's system of governance has evolved into a complex, overlapping set of structures sometimes referred to as the "four pillars" (and occasionally called something rather less complimentary). The European Commission, the Council of Ministers, the European Council, and the European Parliament all have (or had) roles that were defined and redefined—the last time through the Lisbon Treaty drawn up in 2007 and put into effect in 2009. But for all the discussions and deliberations taking place in the EU capital of Brussels, the key decisions are made when the elected leaders of the member states meet. The European Parliament, which has grown from just over 400 parlementarians at its founding to nearly 800 in 2012, remains more symbol than substance and is not a true legislative body.

It is easy to see how dissonance between domestic laws and regulations on the one hand, and EU legislation on the other, creates endless tensions throughout the system. On issues ranging from farming practices and animal rights to immigration and fiscal rules, opinions differ and voters at home tend to resist stipulations made in Brussels. EU bureaucrats are forever tweaking the system to make it work better, and obviously EU governance is a work in progress. At present that edifice does not (yet) look like that of a superpower. But the miracle, given Europe's cultural and social fragmentation and its persistent fractiousness, is that this supranational project has advanced as far and as deeply as it has.

A GEOGRAPHIC PARADOX

Even as European states converge and cooperate on their historic supranational journey, seeking the centripetal ground to cement their unification, other, centrifugal forces drive them apart. To assess the magnitude of the challenge still ahead, it is useful to see Europe's hierarchy of political entities in a seven-rank perspective (Table 9-1). The 28-member, still-expanding European Union sits atop this political-geographical ladder; such still-troublesome entities as Gibraltar, Kaliningrad, Northern Cyprus, and Ceuta and Melilla rank lowest. Between these extremes of success and failure lie five levels of formal and informal power and jurisdiction, all of which the EU must eventually accommodate.

The size and diversity of the states now comprising the EU are greater than ever, and any map showing the "new" Europe as extending from Ireland to Cyprus and from Estonia to Portugal conceals the range of economic, social, and political conditions now incorporated under the EU banner. Inevitably this leads to an "in-group" of leaders and an "out-

TABLE 9-1. EUROPE'S POLITICAL ENTITIES

ENTITY	WHO THEY ARE	OTHERS
European Union	The 25	
Leaders (Core)	France, Germany	U.K., Spain
Followers (Periphery)	Poland, Slovenia, Hungary, Czech Republic	Baltics, Malta, Cyprus
Outsiders	Norway, Switzerland, Iceland	Ukraine, Serbia
Regions	Baden-Württemberg, Lombardy, Rhône-Alpes	Andalusia, Bretagne, Saxony, Tuscany
Devolutionary Pressures	Catalunia, Basque Country, Corsica, Kosovo	Scotland, Flanders, Montenegro
Fragments of History	Gibraltar, Kaliningrad	Andorra, San Marino, Liechtenstein, Monaco, Ceuta and Melilla

group" of followers, a core of original states whose cooperation has advanced furthest and a periphery of countries not able to meet the same criteria of membership (Rachman, 2004). This core, however, shows its own cracks: the map might suggest that France, Germany, Benelux, and the United Kingdom should form all or most of the in-group, but in fact the British have been sufficiently ambivalent about EU participation that France and Germany have become the Union's driving force. The Schengen Agreement, for example, an early five-country multilateral treaty to drop border formalities and ease travel restrictions, included Germany and France but not the United Kingdom.

The power of the core group of states vis à vis the latecomers was evident in 2003 and 2004, when both Germany and France failed to adhere to the economic rules (in context of the growth and stability pact, limiting national debt to 3 percent of GDP) but avoided—in fact, simply voided—the associated penalties. This enraged not only the latecomers but also smaller, less powerful charter members such as the Dutch. There is no doubt about it: the EU is driven by the formidable insiders (Kagan, 2004).

The third tier consists of states that are not members of the EU, including several with strong links to the organization and others less-well connected. The former, Switzerland, Norway, and Iceland, are all qualified, adhere to some (not all) EU regulations, and are involved in the project in many ways. The latter form an outer periphery from sclerotic Belarus and Moldova to aspirant Serbia and Bosnia.

As Figure 9-3 indicates, European countries—EU members as well as non-members—cope with devolutionary forces, some peaceful and historic, others intermittently violent and current. Sometimes opposition to the EU initiative is especially strong in areas where devolutionary forces are particularly strong. The response in the capitals of states affected by such dissension is to "devolve" power to their regions. France, for example, is decentralizing its old and historic administrative framework, dating from Napoleonic times, into 22 historically significant provinces officially called *regions*. These regions (for example Rhône-Alpes, centered on the city of Lyon), though still represented in the Paris government, have their own governments and quite a lot of autonomy in such areas as economic policy and taxation. Importantly, they are also sanctioned by the EU. In Spain, they call that country's 17 divisions *autonomous communities*, and at least one of them, Catalunia, capital Barcelona, has the trappings of a more-or-less independent country, with its own

Catalan language. Germany now has 16 *Länder*, or States. Italy's 20 regions create divergent attitudes in the richer north and the poorer south. In the United Kingdom, there is an especially interesting geographic picture: there are nine EU-sanctioned regions (including London), each of which could have its own assembly; but the U.K.'s historic divisions into England, Wales, Scotland and Northern Ireland simply overpowered these "modern" constructs.

To be sure, the decentralization of administrative frameworks marking the European map is more than merely a matter of politics and administration. When national governments recognize subnational entities such as regions, States, or "autonomous communities," they devolve certain powers to those entities, including budgetary powers. And when the budgets of those regions entail deficits and debts, reflecting lavish spending on local services and projects and sending less money back to the capital, those debts accrue where the euro (like the buck) stops—in the capital. In Spain's case, burgeoning national indebtedness and a widening deficit result in significant part from "hidden" debts run up by the regions, but the central government does not have adequate control over the finances of the "autonomous communities" to rein them in. This, in turn, forces the European Union to step up its oversight of national budgets—which is likely to lead to less "autonomy" for those free-spending communities and regions. This could lead to still another round of devolutionary pressures, even outright campaigns for secession. In early 2012, it appeared that Scotland's voters might soon face a referendum on independence. Although the issues behind this development involve far more than budgetary and fiscal factors, this could be poor timing for the EU; Scotland's secession from the United Kingdom has the potential to revive other, dormant campaigns.

In Table 9-1, it is therefore sensible to differentiate between "purely" administrative divisions of EU states (Regions) and those generating devolutionary pressures with which central governments have to cope. The geographic paradox is that even as Europe is unifying in one direction, it is fragmenting in the other, the two opposites creating still another set of tensions. And then there are Europe's unique fragments of history, exclaves of former world powers or microstates resulting from unique combinations of history and geography. All of these will have to be accommodated in some way as the EU matures; none can threaten the project but some may resist incorporation. The Lisbon Treaty will not be the last word in EU construction.

EUROPE AND THE EURO

Is the EU destined to be a global superpower or a paper tiger? So contradictory are the signals that all forecasts are hazardous. On the one hand, NATO action, with the United States playing a subdued role, was crucial in ending dictatorial rule in Libya in 2011. On the other, NATO's role in Afghanistan remained below expectations. On the surface, the EU remains an unmatched economic success story; it is America's largest trading partner by far. But deeper down, there are problems, with the EU model derided by nationalists at one end of the political spectrum and doubted by political parties governing member states at the other end. The problem always was that the EU initiative was pushed by those seen at the grass roots as an aloof elite, a bunch of know-better bureaucrats out of touch with ordinary citizens.

As long as the economy kept doing well, such doubts could be countered by evidence of well-being: jobs, good salaries, benefits. But then came the economic crisis starting in 2008 and deepening in 2011 with serious financial problems in Ireland, Iceland and Greece and others looming in Spain, Portugal and Italy. Europe came to be seen as being in economic as well as demographic decline, its tax collections inadequate, its pensions unaffordable, its welfare costs out of control, its vulnerable members' foreign debt rising. Euro-members to varying degrees having broken the few rules of euro-participation, the EU now needed more than a Central Bank (already in place): it required a Brussels-based treasury and a finance secretary (or minister) with the power to intervene in and override national budgets.

Toward the end of 2011 the stresses on the EU and the euro project threatened to reverse much of what had been achieved over the previous half century. Two decades after the signing of the Maastricht Treaty and less than ten years after the introduction of the euro, European leaders met in Brussels to draw up yet another treaty, this one to cope with a euro close to collapse and an EU system in urgent need of restructuring. At least there was a sense of reality: no longer was this a matter for the European Commission to address. The real power in Europe continues to lie with the elected leaders of the member states, and they were the ones negotiating a new pact. And the strongest forces were those of Germany and France, respectively led at the time by austerity-favoring Angela Merkel and Nicolas Sarkozy and his notions of "economic-government" run by the "real" powers.

But prime minister David Cameron, representing British interests, refused to agree to the revision of the European Union treaties as proposed, a refusal based primarily on the issue of safeguards for Britain's crucial financial-services industry: EU regulations would, in his view, have endangered the City of London's primacy, including a European financial-transaction tax that the media in England described as a "bullet aimed at the heart of London." After an all-night debate, Cameron issued a veto and went home, where public opinion strongly favored his decision. The other 26 EU members agreed to continue negotiating a new treaty without Britain's involvement, but in truth the chances of success were dim. The United Kingdom, of course, never did adopt the euro, so in that respect its absence would be felt less strongly. But many observers all across Europe warned that unless the euro were rescued by collective and incisive action, and soon, the entire European-Union project might collapse.

In truth, the European Union needs more than a rescue of the euro: it needs persuasive and effective transnational leadership. In the United Kingdom, a majority of voters, according to polls, would vote against EU membership if the matter were put to a referendum, a majority that was never persuaded otherwise even by the previous pro-EU Labor government. Substantial minorities in other EU member states agitate loudly against Brussels. The crisis erupting in 2011 was not just financial: it also was a crisis of governance. When the draft Constitution was rejected by voters in France and the Netherlands, the writing was on the EU's post-Maastricht wall—but leaders failed to heed the warnings.

Today the EU project is seen by too many Europeans as globalization in disguise and unwanted interference in national (and local) ways of life. No matter that, EU or not, Europeans would have been unable to continue retiring too early, receiving pensions and welfare based on outdated fiscal models, enjoying short workdays and long vacations—all without the associated and inevitable sacrifices. It will take better communication and explanation between leaders and citizens than Europeans saw in the heady days of growth and prosperity. In this respect, Europe remains a paper tiger.

10

RUSSIA: TROUBLE ON THE EASTERN FRONT

IN THIS FAST-CHANGING WORLD, WHAT A DIFFERENCE A DECADE makes. In the late 1980s the Soviet Union still was one of the world's two superpowers, a colonial empire extending from the Baltic Sea to Central Asia, a communist enforcer in control of most of Eastern Europe, a nuclear-armed behemoth capable of global destruction. Ten years later, its empire disintegrated, its ideology discounted, and its army in disarray, Russia, the imperial cornerstone, was struggling to reorganize as a democracy and to reestablish a position of consequence on the world's geopolitical stage. But by the middle of the first decade of the new century, Russia's major contest was not with other giants on that stage, but with tiny Chechnya within its own borders. What remained of its armed forces were not at war in some remote Asian frontier but inside Russia itself. Thousands of Russians had died violently, many in terrorist attacks in the capital, Moscow. The cost of this tragedy far exceeded the lives lost and property destroyed. It also compromised Russians' efforts to sustain their march toward democracy, openness, and the rule of law, and brought widespread fears of a return to the authoritarianism that had marked Russian and Soviet governance for so long. Yet a Russia with representative government, whose armed forces are under civilian control and whose laws function effectively, is key to the stability and future economic and political integration of Eurasia.

GEOGRAPHIC PROBLEMS OF A TERRITORIAL GIANT

Not only is Russia the world's largest country territorially: it has more neighbors than any other state. Geographically, nothing is simple when it comes to Russia, and so it is with this set of neighbors (Fig. 10-1). By virtue of its exclave of Kaliningrad, Russia has Poland and Lithuania as European neighbors, as well as Finland, Estonia, Latvia, Belarus, and Ukraine. That makes seven neighbors in Europe alone, and Russia has issues with almost all of them. In the case of Lithuania, Russia wants free transit for Russian freight and military traffic (and no inspections) to Kaliningrad. In Estonia and Latvia, which, like Lithuania are former parts of the Soviet empire, Russia pressed for official status for the Russian language, spoken by large minorities who moved here from Russia during communist times. In Belarus the situation is unusual: that country's Soviet-style, authoritarian ruler has repeatedly pressed Moscow for closer association, even formal union, between the two countries, but the Russians were slow to accept the invitation. With a large Russian minority in Ukraine, Moscow has meddled dangerously in the politics of that large and culturally divided country, once the Soviet Union's second-ranking component in terms of population as well as economic output. Today, many Ukrainian leaders want to see their country join the European Union, but others see their future in closer ties with Russia.

Along its southern border between the Black and the Caspian Seas, Russia borders two former dependencies, Georgia and Azerbaijan (Fig. 10-1). But that simple statement belies a complicated geographic situation that is at the root of many of present-day Russia's most dangerous problems. As a more detailed map shows, Russia's borderland in fact consists of a tier of internal "republics" designated to recognize the non-Russian ethnic composition of their populations, and it is these republics, including Chechnya, that border Georgia and Azerbaijan (Fig. 10-2). This is the region Russians refer to as Transcaucasia, and here live ethnic minorities with memories of Russian subjugation and oppression who would have wanted the same independence given to Georgia and Azerbaijan (and their neighbor Armenia) when the Soviet Union collapsed. But that was not to be, and as a result this region became a cauldron of conflict involving not only the Muslims of Chechnya, who oppose Moscow, but also those of Ingushetiya, who have tried to avoid taking sides, the North Ossetians, who generally support Russia, the Balkars of Kabardino-Balkaria, accused by Stalin of pro-nazi sympathies during

World War II and exiled en masse, and literally dozens of other ethnic groups with turbulent histories. Meanwhile, across the border in Georgia, Russia recognized a province called Abkhazia as independent, and sent an armed force to protect pro-Russian South Ossetia against the Georgian government. And while oil continues to flow from Azerbaijan to Russian terminals on the Black Sea, pipelines are being laid to divert much of it via Armenia to the Mediterranean coast of Turkey. Transcaucasia makes Russia's other borders look uncomplicated by comparison.

Russia's four Asian neighbors also form a contentious group: Kazakhstan, Mongolia, China, and North Korea (Fig. 10-1). When the Soviet Union collapsed and Kazakhstan became an independent state, several million Russians found themselves on the wrong side of the border in northern Kazakhstan, where Russia's space port and launching facilities are also located. The new Kazakh government negotiated agreements to allow these operations to continue, but the Russified north presented far more difficult problems because this area had, in effect, become part of the Russian sphere (as the transport systems on the map confirm). While many Russians emigrated back to Russia, others agitated for secession, prompting the Kazakh government to move its capital from the Kazakh heartland in the southeast to Astana in the Russified north, underscoring its claim to the entire country. Immediately to the east lies Mongolia, a Soviet satellite during the communist period, then closely associated with Russia after 1990 (Russian was the main foreign language spoken here) but now reorienting toward China, currently its largest trading partner by far. This is changing the significance of the Russian-Mongolian boundary, once merely an administrative device but potentially a marker between Russian and Chinese spheres of influence. Still farther to the east, Russia shares a lengthy and historically contentious boundary with China that has been the scene of territorial disputes and cross-border skirmishes. These issues were settled in recent years through negotiations between the two governments, now on better terms. But another issue is emerging: massive cross-border migration by Chinese traders and workers into the Russian Far East (Davis, 2002). And in the farthest reaches of its eastern frontier, Russia has a short but consequential border with North Korea. North Koreans are escaping their tyrannical rulers by crossing into China, and from there some are reaching Russian soil. Indeed, the local regional government wants to encourage this immigration because, as we will see later, Russia's Far East is losing population. But Russia, once North Korea's ideological ally, has other worries,

Fig. 10-1. Russia is the world's largest territorial state by far, but settlement is strongly concentrated in its western, Moscow-centered core and Russian leaders worry that the depopulating east is on the verge of becoming ungovernable.

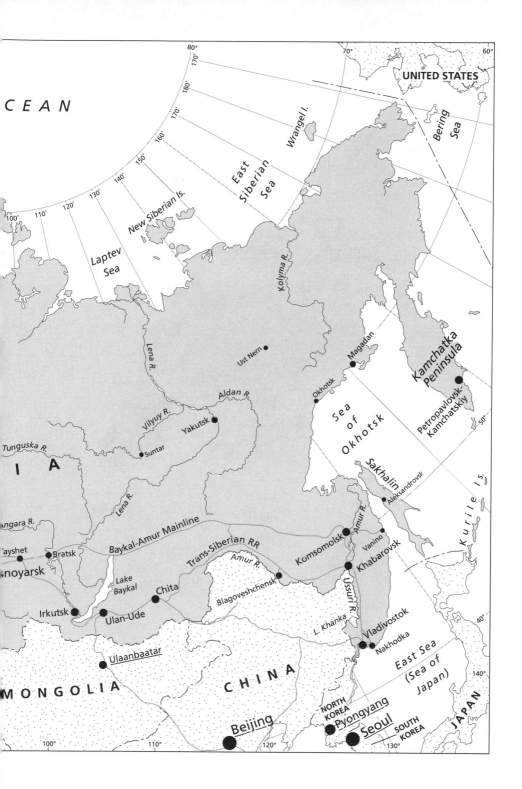

Fig. 10-2. Southern Russia's cauldron of conflict is a mosaic of ethnic republics where Muslim extremists continue to resist Moscow's incorporation and where cultural differences involving other customs and beliefs cause never-ending strife. Across Russia's southern border lie three former Soviet Socialist Republics among which Georgia has seen Russian armed forces invade on behalf of pro-Russian minority Ossetians. Abkhazia, a Georgian province on the Black Sea coast, has been wrested from Georgia by Russia, which recognized it as an independent state.

lying as it does within reach of North Korea's rockets, and potentially, nuclear weapons. Hence Russia is one of the six members of the team of nations seeking to temper North Korea's nuclear ambitions.

Russia's boundaries enclose a country that, for all its continental size, is very nearly landlocked. The czars of old were in a constant drive to push Russia's limits seaward. Peter the Great wanted to make Russia a maritime as well as a land power and built St. Petersburg, his new capital, on Russia's window to the Baltic Sea between Finland and Estonia. Catherine the Great sent her armies to the shores of the Black Sea and into Transcaucasia, but her real objective was an outlet on the shores of the Indian Ocean. The British and the Turks denied her that goal, so that Russia west of the Urals depended on the seasonally ice-blocked Baltic and the narrow, Turk-controlled Bosporus and Dardanelles for outlets to the sea. True, Russia's eastward expansion to the Pacific Ocean gave it a major port at Vladivostok, but this was no practical alternative to a nation concentrated thousands of miles across Siberia to the west. And Russia's borders create another problem obvious from the map: for all its bulk, the country is almost entirely confined to high, cold latitudes under Arctic influences much of the year (see Figure 5-3). Grain shortages during the Soviet era drove communist planners to expand farm production through irrigated megaprojects in the republics, but even then Moscow had to depend on costly imports from the west. Now the Soviet Union's breadbasket, Ukraine, is an independent country and Russia's climatic quandary is even more pronounced. As a Russian geographer once said to me, "our borders have never been our friends."

Crossing those borders overland during the Soviet era was a daunting experience. I rode a bus from Helsinki to (then) Leningrad in June 1964, my first field trip to the U.S.S.R., and I learned that the boundary on the map was in fact a wide zone on the ground. We reached the Finnish border just before dark, and formalities were quick and courteous. We reboarded the bus and proceeded into a treeless corridor, to be met by a carload of uniformed, armed guards who escorted us several miles down the road. In the distance was a patch of bright light, so bright that when we reached it, it was as if the bus had been driven into a surgical theater. There we were ordered off the bus and all luggage and cargo was unloaded. Passengers were separated into three groups, Soviet, European (I was traveling on a Dutch passport), and others, including a group of Canadian and American academics. Every piece of luggage was examined in minute detail, and we were physically searched in ways that make the current airport-security procedure seem casual by comparison. Then we were instructed to sign documents stipulating that we were not carrying items ranging from books and "documents" to weapons and "propaganda." The entire operation took about three hours, and I wondered how long the wait would be when a line formed. "Never a queue," said the English-speaking guard handling the North Americans. "Only three buses a day and maybe five cars." I thought about that as heavy armored gates swung open and our driver headed into the darkness of the road to Vyborg. The main road between the capital of a neighboring country and the main port of Russia, and the daily traffic amounted to fewer than ten vehicles. Soviet borders were barriers indeed.

A VAST REALM

Even after the loss of its 14 dependencies, Russia remains the world's giant state territorially, nearly twice as large as the next-ranking country, Canada, and with 13 neighbors (no. 2 Canada has one). From the volcano-studded Kamchatka Peninsula in the Russian Far East to the great port city of St. Petersburg in the west, the country stretches across 11 time zones, artificially reduced to nine in 2009. A television program called *Good Morning Russia* airing in Moscow at 7 a.m. local time would be seen in Vladivostok near dinner time. Russia's northernmost Arctic-sea islands lie north of 80 degrees latitude; its southernmost sliver of land adjoins Azerbaijan in Transcaucasia, still above 40 degrees. Putting this in North American terms, all of Russia lies north of the approximate latitude of Boston.

It is worth taking a moment to look at the map of world climates (page 143) to see just how cold Russia really is. Neither distance nor mountains protect it against invasions of Arctic air. Almost all of it is dominated by D climates, which in the west have a short warm summer that diminishes eastward, creating the frigid conditions for which Siberia is a synonym. Not until maritime influences moderate the climate along the Pacific coast do Siberian conditions let up. Looking at the map of world population distribution (page 83) we can see that most of Russia's over 140 million people cluster in the mildest corner of the country, the west, and in a ribbon along the southern margins of Siberia, where the Trans-Siberian Railroad links cities and towns and connects the populated west to the sparsely peopled Far East.

GLOBAL WARMING: HOPE FOR RUSSIA?

As Figure 10-1 shows, Russia's coastline lies entirely on the polar side of the Arctic Circle, where the climate is frigid. This lengthy coastline has not been an advantage to the Russians; most of the Arctic Ocean is frozen much of the year, and only the warmer water of the North Atlantic Drift, looping around northern Scandinavia and Finland, keeps the ports of Murmansk and Arkhangelsk open a few weeks longer.

But the Russians may get help from climate change. If, as projections indicate, the current cycle of global warming continues, several crucial environmental changes, some but not all of them favorable, may bring a new era to the Russian state. In the first place, the Arctic Ocean's winter ice will shrink, keeping ports open longer, opening up a Russian Maritime Passage along the northern coast that may alter global shipping routes. Secondly, continued melting of the ice will enable exploitation of huge energy resources known to lie beneath the waters of the Arctic Ocean. Already, the Russians are staking claims to the Arctic's continental shelf and, in a bit of high drama, a Russian mini-submarine in 2007 planted a metal Russian flag at the North Pole beneath the Arctic ice, proclaiming Russian intentions. Thirdly, high-latitude warming will melt much of the permafrost (permanently frozen ground) of the far north, possibly leading to northward expansion of the forest and shrinkage of the barren tundra.

Optimistic predictions suggested that warmer, moister air masses might also improve farming on the Russian Plain, but as always, nature has a way of confounding the forecasters. In 2010 Russia experienced its hottest summer on record, but the heat wave was accompanied by drought, not moisture. Forest fires broke out all over the central and southern areas, destroying thousands of villages, killing nearly 60 people and ruining crops. The grain harvest, 100 million tons in 2009, was down to 60 million in 2010; grain prices soared, and the government banned grain exports to hold local prices down. Meanwhile Moscow was shrouded for weeks in a poisonous smog that was a daily reminder of nature's power to bewilder the experts.

THE GREAT REGIONS

Just as Americans use geographic references such as "Midwest" and "Great Plains," so do Russians refer to their vast country's broad physiographic regions. The great divider of Russia is the Ural Mountains, the Appalachians of Russia but located much farther into the interior (Fig. 10-3). The Urals extend from the Arctic Ocean, where they rise above the water as glacier-carrying Novaya Zemlya Island, to (and beyond) the desert border with Kazakhstan. West of the Urals, in the perception of many people, lies "European Russia"; to the east, therefore, lies something else, though the cultural landscapes of Russian towns to the east of the Urals are remarkably similar to those of the west. In any case, the heart of Russia, its core area in geographic lingo, lies on the Russian Plain, an extension of the North European Lowland, cooler and drier but still productive agriculturally. At the center of it is Moscow, on its short Baltic coast lies St. Petersburg, and crossing it is the great Volga River, flanked by old industrial cities all along its course.

Siberia begins on the eastern slopes of the Urals and does not end until the shores of the Bering Sea, but its relief does change from west to east (Fig. 10-3). Westernmost Siberia, region (3) on the map, has comparatively low relief and is drained by a major river system, the Ob-Irtysh, whose gradient is so slight that Soviet engineers talked of reversing it to irrigate farmlands to the south (Lincoln, 1994). In this forbidding, forested, frigid countryside lay many of the prison camps of the infamous Soviet gulag, in which, historians estimate, between 30 and 60 million inmates perished during the seven decades of communist rule (Remnick, 1993). Along Siberia's more livable southern margin lie cities such as Omsk and Novosibirsk, strategically crucial during World War II when much of Soviet industry was shifted eastward, across the Urals and away from the nazi advance.

At the eastern margins of the West Siberian Plain, the relief changes quite dramatically, especially in the south, where jumbled mountain ranges rise from the plain. In the north, Siberia takes on the character of a rugged plateau. Here the Trans-Siberian Railroad passes through narrow valleys and hugs the walls of steep gorges, eventually emerging from this rough terrain to reach the key city in the area, Irkutsk, gateway to Lake Baykal. This freshwater lake lies in a rift valley similar to those of East Africa's Great Lakes, but Lake Baykal, nearly 640 kilometers (400 miles) long and averaging 50 kilometers (30 miles) in width, is even deeper, reaching more than 1,620 meters (5,300 feet) in depth. By some calculations Lake Baykal contains one-fifth of all the liquid freshwater on the Earth's surface, and its unique ecology attracts an endless stream of researchers from all over the world to study it.

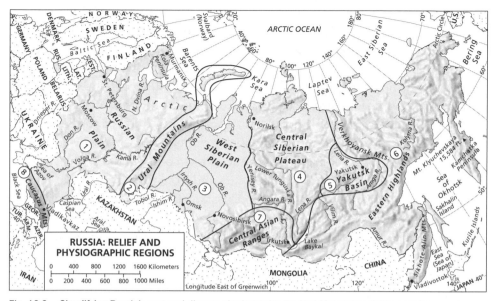

Fig. 10-3. Simplifying Russia's vast and diverse physiography: the Ural Mountains divide west from east, where Siberia dominates and settlement forms a discontinuous ribbon along the mountainous south.

Now comes Russia's vast, forested, mountainous east, region (6) on the map, lower in the Yakutsk Basin and higher in the spectacular Kamchatka Peninsula, the country's most geologically active zone. Don't expect to drive to this earthquake-prone, volcanic slab of tectonic plate (ironically, northeastern Russia is geologically part of the North American Plate!), because there are no connecting roads. The people who share this peninsula with more than 20 active and over 100 dormant volcanoes live as though they were on an island, fishing for a living and boating or flying to the mainland when the need arises.

As Figure 10-3 shows, the Russian Far East incorporates one real island, named Sakhalin, and this is an important component of this region's physical as well as cultural geography. From the mid-nineteenth century on, the Russians and the Japanese repeatedly fought over Sakhalin Island, and not until the end of World War II was Soviet control confirmed. When the Russians held it, they used Sakhalin as a penal colony (the great writer Anton Chekhov in one of his books described the terrible conditions under which prisoners lived), but during Soviet times Sakhalin became an increasingly important source of fuels ranging from oil in the north to coal in the south. In post-Soviet years additional finds of oil reserves have made Sakhalin Island a key constituent of the commodity-based Russian economy.

Russia's enormous size bestows it with a large inventory of natural resources, among which oil and natural gas have been the key money makers during the post-1991 period. The Soviet Union's first dictator, V. I. Lenin, was determined to speed his country's industrialization, especially its heavy manufacturing. For this the U.S.S.R. contained almost everything it needed, from coal to iron ore and from other metals to alloys. When World War II loomed, this resource base allowed the Soviets to build their own weaponry with which to defeat the German invaders, and afterward Russian factories produced most of what the country required, from automobiles to railroad cars and from tractors to passenger planes. When, during the 1950s, the Russians launched the first Earth-orbiting satellite and a Russian cosmonaut was the first in space, it was a homegrown project that astonished the world, proving, it seemed, the superiority of the Soviet system.

But for all its size and large population (third largest in the world while the U.S.S.R. lasted), economic interaction with the rest of the world was very limited. Soviet products did not appear on international markets; Russian automobiles were not seen, except as a curiosity, on foreign streets. The state enterprises of the command economy produced goods at costs and quality levels that would have made them uncompetitive in any case, so that the largest volume of exports was not consumer goods but weaponry, sold by the Soviet government to allied regimes. The economic geography of the Soviet Union resulted from assignment, not efficiency—certain cities and towns were allocated production tasks based on criteria other than cost. The potential for corruption in such a system can only be imagined. When the Soviet Union collapsed and Russia faced integration into the world economy, the massive sale of commodities—oil and natural gas—became the major source of badly needed external revenues. Russia was fortunate to possess substantial energy reserves from its share of the Caspian Basin in the west to Sakhalin Island in the east, with ready customers including Europe, Japan, and China. But overdependence on a single commodity entails risks and skews development. It opened the way to "crony capitalism" as high-level insiders bought up state enterprises, tax collections faltered, factories failed, and, in 1998, the government defaulted and the economy crashed, driving away the first wave of investors. A vast reservoir of natural resources alone is not enough to ensure this (or any other) country's prosperity.

Today, nevertheless, Russia is often called a "petro-ruble state," a country almost as heavily dependent on the export of energy resources as Saudi Arabia, the "petro-riyal state." Yes, there are trappings of a modern market economy: private companies, private property, foreign investment, stock markets. A class of ultra-rich built skyscrapers and live in mansions, buy foreign assets (such as football clubs) and send their children to private schools. But for millions of ordinary citizens, the transition has been wrenching because social institutions failed, the legal system faltered, corruption mushroomed, organized crime flourished, and the domestic economy (other than that linked to the energy industry) struggled. True, Russia faced—and continues to confront—unusual problems arising from its vast dimensions, scattered and culturally diverse population, and communist economic and administrative legacies. But it has also suffered from failures of leadership and inadequately representative government. Its democratic shortfall is tragically represented by the predetermined Putin-Medvedev "leadership" alternation, as though Russians must permanently endure the absence of free and competitive elections in which issues, not personalities, dominate.

SOVIET LEGACY, RUSSIAN CHALLENGE

In the political-geographic arena, too, the problems Russia confronts are daunting. Organizing the administration of so vast, remote, and isolated a realm has posed a historical challenge for Russian czars, communists, and democratically elected leaders alike. Expansion was one thing—Russian armies were able to penetrate deep into Central Asia, and Russian colonists claimed Alaska and built their southernmost fort near San Francisco Bay—but consolidation was quite another. When United States secretary of state William Seward offered to buy Russia's Alaskan holdings in 1867, the Russian government quickly agreed, because these outposts were becoming more trouble than they seemed to be worth. In truth, successive czarist rulers never established a satisfactory administrative structure for the numerous peoples, Russian and non-Russian, under their control. Europe's democratic revolution passed Russia by, and its economic revolution touched the czars' domain only slightly. Most Russians, and tens of millions of non-Russians under czarist domination, faced exploitation, corruption, subjugation, and starvation. When desperate rebellions erupted in 1905 and full-scale revolution broke out in 1917, there was no political framework to hold the state together (Shaw, 1999).

Thus it fell to the communist victors in that revolution, the Bolsheviks, to design the regional framework that would constitute the Soviet Union. The basic structure created 15 "Soviet Socialist Republics" (SSRs) among which the Russian Soviet Federative Socialist Republic (RSFSR) was first among theoretical equals. The other 14, from the Estonian SSR on the Baltic to the Tajik SSR bordering Afghanistan, were designed to accommodate non-Russian peoples who had fallen under Russian domination during czarist times. That system, of course, fell apart when the Soviet Union collapsed. More durable was the framework the communist planners laid out within Russia itself, because this is what the post-Soviet leaders inherited, and with which they had to work in their effort to transform the vast country from communist dictatorship to democratic consensus.

It is useful to take a look at this convoluted Soviet system, because it contains the seeds of the troubles post-Soviet Russia has faced in the years following 1991 (Shaw, 1999). The Soviets, always mindful of status and hierarchy, divided the RSFSR into internal "republics," autonomous regions (*okrugs*), provinces (*oblasts*), and territories (*krays*). The "republics," like those beyond Russia's borders, were established to recognize the largest ethnic minorities within the RFSFR. Although this administrative order reflected a descending level of importance and, in a very general way, distance from the capital, that order could be countermanded through the powerful personality of a local leader. Kremlin watchers knew that when you began to hear the name of some remote kray frequently, it was likely that a party leader from there was ascending the political ladder and gaining influence in Moscow. Nor were all krays unimportant by definition: the one located farthest from Moscow, Primorskiy Territory, was home to the huge Soviet naval base of Vladivostok, a strategic city closed to the outside world whose borders were controlled even more tightly than those of the country as a whole.

The Russian "federation," of course, was a federal state in name only. The RFSFR, like the Soviet Union as a whole, functioned as a centralized unitary state, and all essential power resided in Moscow. The rights of ethnic minorities, despite their "republics" on the map, were strictly limited, and during World War II minorities tended to be suspect as potential allies of the invading Germans. The story of what happened to the Muslim Chechens is among the worst: Stalin accused them of collaboration and in 1944 ordered the entire population loaded on trains and exiled to Central Asia. Tens of thousands died along the way, their bodies thrown from the railroad cars. Many who survived this horror then perished in the harsh and unfamiliar environment at their destination. A man named Shamil Basayev claimed to have lost 40 relatives in this genocide, a fateful as well as dreadful personal calamity. Pardoned by Stalin's successor and permitted to return to their homeland in 1957, the remaining Chechens never forgot what Russians did to them, and when the Soviet Union collapsed in 1991 they seized on the opportunity to declare their independence. This started a cycle of violence that killed thousands more and continues to this day—not just in Chechnya but in Moscow and elsewhere.

When Russia emerged from the wreckage of the U.S.S.R. as a geographically redefined country ostensibly embarked on a course toward democratic government and a true federal system, Chechnya was not its only problem. The new Russian administration, led by the redoubtable Boris Yeltsin, could not simply sweep away the structural legacy the Soviets had built; it was the only game in town. So the Russian leadership took stock of the Soviet map and began to modify it to facilitate what would be a difficult transition. Counting all the administrative entities (whatever their rank) in the Soviet system, Russia was endowed with 89 regions, including 21 ethnic republics. The new government decided to give all regions an equal voice in the government, although special status was given to the republics. For some of these republics, however, this was not enough. Early on, Chechnya's Muslim inhabitants proclaimed their desire for self-determination; also-Muslim Tatarstan, astride the Volga east of Moscow, went so far as to fly its own flag, to launch its own airline, and (like Chechnya) to refuse to sign the Russian Federation Treaty.

TROUBLE IN TRANSCAUCASIA

In time, common sense prevailed, and most of the would-be rebels came in from the cold. But not the Chechens. The old Soviet map had a joint ethnic republic for the Chechens and their Muslim neighbors, the Ingush, but they never did get along, so in 1992 the Russian parliament approved a split: Ingushetiya would be home to the relatively cooperative Ingush, and Chechnya would house the militant Chechens (Fig. 10-2). Almost immediately, the Chechens' leaders declared independence,

Fig. 10-4. The mouse that roared: Chechnya on
the northern flank of the Caucasus Mountains.

starting a cycle of increasing violence that would resonate in Moscow
itself.

About the size of a small New England State and lying against the
north slope of the Caucasus Mountains between the Black and the Caspian Seas, Chechnya has three landscapes: the Caucasus mountains to
the south, a historic refuge for Chechen rebels; the plains to the north of
the Terek River, where Russians, who make up about 30 percent of the
population, have been farming for generations; and, between these, the
urban-industrial middle zone where the capital, Groznyy, other towns,
and the territory's oil installations are located (Fig. 10-4). Power over
this middle zone means control over the territory, and here most of the
conflicts that began in 1994 have been fought as the Russian army, at
times exceeding 80,000 soldiers, engaged rebels in a war conjuring up
memories of Afghanistan, or perhaps Iraq. The capital was totally devastated, other towns were severely damaged, and negotiations after stale-
mates soon were followed by renewed fighting, with high casualties on
all sides. Meanwhile what had been a nationalist campaign for independence turned into a wider war involving Islamic causes as fighters and
funds from Afghanistan to Saudi Arabia arrived to support the Muslim
Chechens (Tishkov, 2004).

However, Chechnya is only one of a tier of seven ethnic republics along the northern flank of the Caucasus mountains, from restive and violent Dagestan in the east to calmer Adygeya in the west (Fig. 10-2). Each one of these republics presents its own particular set of cultural, political, and economic problems in this fractious region: Dagestan's 2 million people are divided into some 30 ethnic groups; mainly Muslim Ingushetiya is divided between pro-Russian and pro-Chechen supporters; North Ossetians, generally pro-Russian, want union with their South Ossetian neighbors across the border with Georgia; the Muslim Balkars in Kabardino-Balkariya have memories of Russian mistreatment during World War II; many Muslim Karachay in the Karachayevo-Cherkessiya Republic were forcibly exiled during that war as well; and only Adygeya is without latent ethnic conflicts. Sympathies for the Chechen cause extend well beyond Chechnya's borders, even after terrorists began to extend the conflict into this region and beyond.

This terrorist campaign, which soon reached Moscow itself, did incalculable damage to the Russian state and its political and economic hopes. Shamil Basayev, the Chechen who lost so many family members during the Chechens' Central Asian exile, was in Moscow supporting President Boris Yeltsin during the 1990s, seeing the president as Chechnya's only hope for autonomy. But when that hope was dashed with the 1994 intervention by Russian forces, Basayev was ready to play his role. In 1995, after Russian attacks leveled his family's home and killed 11 of his relatives including his wife, two daughters, and a brother, Basayev became Russia's Usama bin Laden. He began a terrorist campaign that continued for more than a decade, unsettling not only the region around Chechnya but also Moscow itself.

Basayev's first high-profile action occurred a month after his family's demise, an attack on a hospital in neighboring Dagestan in which his fighters took 1,500 hostages killing more than 100. In 1996 his forces drove the Russian army out of Groznyy and in effect achieved the autonomy Moscow had denied him, but in the "presidential" election that followed he received less than a quarter of the vote. Although he was included in the government, Chechnya in effect was a failed state by then, and Islamic jihadists and Arab funds were flowing in. Basayev became prime minister of the Chechen-controlled part of Chechnya they referred to as the Republic of Ichkeria. With Russia in political disarray and Boris Yeltsin about to relinquish his presidency, Basayev's terrorists mounted another raid into Dagestan and bombed two apartment buildings in Moscow. When the new president, Vladimir Putin, ordered Russian forces back into Chechnya in late 1999 for what was to be the second war for control, he had the almost universal support of Russians. The Chechen regime they attacked was recognized by only one other government: Afghanistan's Taliban. By midwinter 2000, Chechen forces had been driven out of devastated Groznyy and into their mountain hideouts, and from then on Basayev was reduced to planning and ordering a series of terrorist strikes. Perhaps the most dramatic was the takeover of a crowded theater during a performance in Moscow in October 2002 by a group of 41 Chechens and their allies, leading to a prolonged standoff, a bungled rescue effort, and the deaths of 130 theatergoers.

Consider the impact on Russia of such actions in just the year 2004: in February, a bomb in the Moscow subway killed 41 and injured more than 100. In May, a bomb planted under a row of seats in a Groznyy stadium killed the new Moscow-approved president, Akhmad Kadyrov. In June, an assault on police stations in neighboring Ingushetiya killed nearly 100. In August, a raid on police installations in Groznyy killed more than 50. In September, a team of terrorists took more than 1,000 children, parents, and teachers hostage in a school in Beslan, North Ossetia, resulting in the deaths of at least 370 children and adults. Also in September, suicide bombers simultaneously blew up two airliners flying from Moscow's airport, killing 90. With no prospect of changing the course of history through such actions, the Chechens, having lost their republic and seen many of their allies depart to fight the infidels in Iraq, now took opportunistic revenge and waited for a time when their cause would return to center stage.

THE PUTIN ERA

The year 2004 marked an important threshold in Russia. President Vladimir Putin, first elected in 2000, was re-elected for another four years and began to consolidate his power in important ways. In his first year in office, Putin had intensified the military campaign in Chechnya with favorable results as most of the Muslim extremists were driven from Chechnya's Middle Zone into the mountains. Now he seized on the public furor over the Beslan school massacre to propose a set of measures that would ostensibly strengthen the Russian state against terrorist threats, but which in fact strengthened his own hand in domestic politics. For example, regional governors would henceforth be appointed by the president rather than elected by the regions' people—essentially putting an end to Boris Yeltsin's efforts to create a real and democratic federation from the communist system that had been a federation in name only.

This and other moves made by Putin were intended to strengthen his personal power, but polls indicated that he had widespread public support. He frequently expressed his concern over the continuing terrorist campaign, which by 2005 had spread into Dagestan and beyond; in Nalchik, the capital of Kabardino-Balkaria, Muslim fanatics killed more than 100 people in a carefully planned attack. Putin's policy was to "Chechenize" the conflict in Chechnya, pitting pro-Russian activists against diehard rebels, a policy that had some effect in confining the action to the republic itself. Shamil Basayev, the terrorist leader who had inflicted so much damage in Russia as well as in Chechnya (including planting a bomb that killed pro-Moscow Chechen president Akhmad Kadyrov), was himself killed when his truck loaded with dynamite headed for an attack in Ingushetiya exploded in July 2006. Although Putin could not take credit for Basayev's demise, it happened on his watch and marked a milestone in the pacification of Chechnya.

Terrorism, however, continued to plague Russia. Apart from attacks in Transcaucasia that mainly involved Muslim and non-Muslim locals, terrorist strikes continued in Russia itself, including the November 2009 bombing of the express train between Moscow and St. Petersburg that killed more than 30 people. In March 2010 suicide bombers from the Transcaucasus set off explosions in two Moscow subway stations, causing 40 deaths. And on January 25, 2011 a bomb exploded at the international arrivals terminal of Moscow's Domodedovo Airport, killing 35. The terrorist threat remains a constant reality in Russia, sowing fear and anger and imperiling the fragile peace among Russian and non-Russian residents.

DEMOGRAPHIC DISASTER

Russia's political, economic, and strategic struggles continue against a background of social problems so severe that they are routinely described by demographers as disastrous (Demko, 1998). In 1991, when reconstituted Russia emerged from the disintegrated Soviet Union, its population was approximately 148 million. By the beginning of 2012, it had declined to just over 141 million—even though several million ethnic Russians had immigrated during this period from the neighboring former Soviet republics. Geographers who study population issues cal-culate that, since the end of communism, Russia has seen about 10 mil-lion more deaths than births. Such population decrease usually accom-panies a lengthy and major war or massive emigration. But in Russia, neither war nor an outflow of people is to blame. Rather, the situation signals severe social dislocation.

It should not be surprising that Russia's birth rate dropped markedly during and after the breakup of the Soviet Union, as uncertainty tends to cause families to have fewer children. Although the birth rate more or less stabilized just below 9 per thousand, the death rate has surged to more than 16 per thousand, causing an annual loss of population of over 0.7 percent or nearly 1 million annually. Only net immigration slowed a decline so severe that many demographers refer to it as catastrophic. For two decades, Russia was in demographic free fall.

What caused this calamity? The birth rate was held down by wide-spread abortion practices and by sexually transmitted diseases, but the death rate, especially among males, revealed the real trouble: rampant diseases such as tuberculosis, heart disease, and underreported AIDS, endemic alcoholism linked to these diseases and also part of a culture of excess in which vodka and more recently beer play a major role, heavy smoking, especially among young males, traffic and industrial accidents, suicide, and murder. On average, a Russian male is nine times more likely to die a violent or accidental death than his European Union coun-terpart. Male life expectancy in Russia declined from 71 in 1991 to 59 in 2004 (female life expectancy also dropped, but much less, to 72). Fewer than half of today's Russian teenage males will reach 60.

How to confront this crisis? The Russian Parliament, the Duma, has been pressing the fast-growing beer industry to limit its advertising and has tried to legislate against beer consumption in the streets, but the culture of vodka consumption is so entrenched that this campaign is unlikely to have the desired effect. Yet if significant population loss con-tinues, Russia may have a mere 100 million inhabitants by 2050, possibly fewer, and given its vast territory, this may so weaken the state as to make it unsustainable. In the two decades 1990 to 2010, the Far East lost 17 percent of its population, the South 12 percent, the Northwest, more than 8 percent and Siberia nearly 5 percent (Fig. 10-5). The Far East, with an area the size of the contiguous United States, now has a mere 6.7 mil-lion inhabitants, labor shortages there are common, and development is mostly stalled (Thornton and Ziegler, 2002). The only solution seems to

RUSSIA: THE NEW FEDERAL DISTRICTS PROCLAIMED MAY 13, 2000, AND MODIFIED IN 2009, AND THEIR CAPITALS

Fig. 10-5. Russia's Eight Federal Districts, proclaimed by President Putin and superimposed on the administrative map of Russia's more than 80 historic regions. Two decades of population loss are recorded in the numbers.

be a large influx of immigrants, and they are ready to come: North Koreans, Chinese, and others. But whether the government in Moscow will approve the immigration of as many as 250,000 East Asian immigrants is another matter. The Russian presence in the Far East is already tenuous, and the arrival of large numbers of Koreans and Chinese might create new social problems even as it begins to solve economic ones. Russia is in demographic trouble, and the way out is not in sight.

The Russian government announced in 2010 that the population decline had leveled off, and that Russia had actually gained about 25,000 citizens in the previous year. Citing its efforts to improve social conditions, the Medvedev administration suggested that the worst was over and President-elect Putin vowed that he would preside over a new, positive demographic era. Don't count on it.

NEW ERA, OLD PROBLEMS

In 2000 Russia achieved a feat that a decade earlier was unimaginable: a democratic transition in the president's office. Vladimir Putin succeeded Boris Yeltsin, who for most of the decade held the office. Younger, more energetic, and determined to set Russia on a new course, he took on the oligarchs who had been favored by his predecessor, revived the economy, and began to reform the armed forces. He made it clear in public statements that he wanted Russia to acquire a stable political system, become a world-class economic power, and regain the international respect and strength it had lost during the nineties. He also proclaimed an adherence to the rule of law, and used this principle to corral several of President Yeltsin's favored oligarchs and put them in jail. But by rule of law, President Putin seemed to mean his own arbitrary power rather than a set of regulations passed by the Duma, the lower house of parliament.

Russia has a history of authoritarian government that goes back centuries, and the decade of President Yeltsin's chaotic and warped versions of democracy and capitalism made many Russians yearn for a stronger leader, even the old autocratic kind (Trenin, 2002). To a certain degree, this is what they got, and the ongoing conflict in Chechnya and its terrorist extension did much to create the opportunity. Immediately after the Beslan atrocity, President Putin said that "democracy does not result in stability, but rather instability . . . it does not unify, but rather divides" (Myers, 2004). Ethnic and religious tensions in culturally divided areas can only be controlled "with an iron hand from above." Most ethnic Russians appeared to agree with him, and many seemed willing to yield personal freedoms to accomplish it. Living in fear makes people worry less about individual rights and freedoms, and skillful exploitation of such an atmosphere can allow political leaders to concentrate power. President Putin approvingly pointed to the absence of Chechnya-like conflicts during the Soviet period, when such strife was "harshly suppressed by the governing ideology." Seizing the opportunity, the president began taking control of still-independent television, radio, and other media, and started manipulating regional and local elections.

The strategy to achieve more effective control over Russia's 89 politically still unpredictable regions was foreshadowed some years ago when the first Putin administration divided the country into seven new administrative units—not to enhance their influence in Moscow but to expand Moscow's authority over them (Fig. 10-5 displays these units, with one later addition, North Caucasus). Each of these "federal" administrative districts has its capital city, which became the conduit for Moscow's "guidance," as the official plan put it. Then, in 2004, the other shoe dropped. President Putin introduced legislation to reverse the fundamental democratic right of representation in the regions, enshrined in the Constitution, by ending the regions' authority to elect its own governors. Henceforth, these governors would be appointed by the president, not elected by the people in the regions. Earlier, the governors had already lost their membership in the Federation Council, the upper house of parliament. Coupled with this move, the president announced plans to change the electoral system for seats in the Duma, so that party appointees, not independent legislators, will fill those seats. All this was approved in the Duma, where parties loyal to the president have a huge majority; it was even supported by some of the elected governors themselves. It was, in the words of a Russian colleague paraphrasing China's line on economics, a retreat to "democracy with a Russian face." Western governments, however, expressed strong reservations.

Concerns over Russia's apparent drift toward a new authoritarianism were reinforced by statements Putin made following his re-election in 2004. At a meeting with President G. W. Bush in 2005, President Putin reiterated his doubts about democracy: if the Russian version was not to America's liking, he cited the Electoral College as evidence that America's democracy had its own contradictions. As to restrictions imposed on Russia's media, this, too, had precedence in other not-so-democratic "democracies."

Putin could not run for another term in 2008, but he had already anointed his successor: Dmitry Medvedev, an aide and assistant who had never run for elected office. Putin's endorsement guaranteed Medvedev the presidency; Putin took the position of prime minister and was the power behind the throne. Almost immediately, there were suspicions within and outside Russia that this arrangement would simply serve to allow Putin to return to the presidency in 2012 while abiding by the two-term limit imposed by the Russian Constitution. And so it turned out. In the autumn of 2011 Putin announced that he and Medvedev had "agreed" to switch offices at the time of the (forgone) 2012 presidential "election." So much for democracy.

Make no mistake: Putin was a very popular president when he was in office, and he would probably have won a free and open election in 2012. He had saved Russia from economic collapse in the 1990s, had tackled the Chechnya issue forcefully and effectively, had benefited from the sharp rise in world energy prices during his first two terms and did much to restore Russians' confidence in their state and revive a long-dormant nationalism. Russia, once again, was a power to be reckoned with, and Putin had made it happen.

RUSSIA IN THE WORLD TODAY

Russia may not have the capacity to recapture the position of global power once held by its Soviet predecessor, but Russia is indeed a force in the modern world. Russia retains a nuclear arsenal whose status and operational condition are uncertain, but its armed forces are no longer in the disarray of the 1990s. Russia has waged war in neighboring Georgia when it saw its interests threatened; it has reacted to Western security initiatives by threatening to place missiles in Kaliningrad aimed at anti-ballistic installations in Europe. Russia maintains military bases in nine of its former 14 Soviet partners. Russia's large and still-growing energy inventory are crucial to Europe and increasingly so to China and Japan. And Russia has taken independent, sometimes obstructionist positions in international strategic matters; in 2011 it joined China in the UN Security Council to veto sanctions against the murderous regime of Bashar al-Assad in Syria. It has given support to Iran in its nuclear ambitions and has supported Serbia in its opposition to independence for Kosovo (in 2012 Russia was still among the minority of states that refused to recognize Kosovo's sovereignty). In November 2011 Russia's then-president Dmitiri Medvedev warned that Russia would deploy its own missiles and could consider withdrawing from the New Start nuclear arms reduction treaty if the United States decided to move forward with its stated plans to install a missile-defense system in Eastern Europe. Although this system was designed to defend against a potential missile attack from Iran and would set up interceptors in Romania and radar support in Turkey, Moscow chose to interpret it as a threat to its own strategic deterrent. Its decision to link the issue to the just-ratified New Start treaty may reveal new doubts about the treaty or perhaps a difference of opinions within the Kremlin. If this combination of tactics appears to be more opportunistic than ideological, it does reflect Moscow's desire to make Russia a more consequential factor in the deliberations that engage the international community. They are also designed to stoke Russian nationalism, always an advantage when it comes to elections.

But the elections held on December 4, 2011, nonetheless produced a series of surprises and, quite suddenly, seemed to alter Russia's political-geographical landscape. In Putin's Russia, electoral fraud had become routine, state television was essentially the propaganda arm of United Russia—the party of Putin and Medvedev—and political monopoly was the price to be paid for stability and prosperity. But this time it turned

WHAT WENT WRONG?

When the Soviet Union fell apart more than two decades ago, hopes were high worldwide that the new Russia emerging from the wreckage would embrace Western values and play a key role in the fast-changing international community.

But before long a wave of nostalgia and a rising tide of disillusionment, coupled with an aversion to disorder and a fear of uncertainty, dashed those hopes and brought a return to an authoritarianism, corruption and bureaucratic monopolism reminiscent of Soviet times. What happened?

Communism's institutional imperatives die hard, and Russia's weak institutions formed a route to bureaucratic power. The police, the courts, the media, the schools, the banks, the arms—all could be (and were) usurped, first by Yeltsin-era officials and oligarchs who assembled fortunes as property owners and energy-industry moguls, next by Putin-era bureaucrats who turned on the oligarchs and officials-turned-businessmen for their own reward. As *The Economist* observed in December 2010, "What Stalin wrought by repression and extermination, today's Russia achieved by corruption and state violence."

Russia's people still associate personalized state power and enforced social stability with a time when the Kremlin was one of the world's two dominant centers of power, and their tolerance for a "mafia state" has endured in part because of rising incomes derived from the state's energy exports. But the signs of a reckoning seem to be at hand.

out to be different. United Russia, which claimed nearly two-thirds of the vote in the previous election, was down to less than 50 percent officially, but allegations of widespread fraud suggested that its true tally was significantly less. The media reported that the polling station where Putin voted, and where oversight was stricter than perhaps anywhere else, United Russia got less than 24 percent and lost to . . . the Communist Party, which got over 26 percent of the vote.

In the absence of genuine democratic opposition, voters showed their discontent by voting for the Communists and for other parties that, as in the case of Yabloko, were not even represented in parliament. And in an unprecedented outpouring of anger, they took to the streets in protest, and were brutally dispersed by riot police; meanwhile the regime sought to censor online social networks carrying denunciations and calls to action. In the media of the international community Putin was so eager to impress, Russia's fraudulent election was denounced and ridiculed (*Newsweek* called Russia the "Incredible Shrinking Superpower" careening toward irrelevance and *The Economist* called on Putin to "clean up the Kremlin and modernize the economy"). Those who anticipated that the spate of protest would be brief, citing Russians' historic preference for stability, were surprised by the estimated 100,000 who streamed to a protest rally in Moscow three weeks after the election, and sporadic demonstrations in Moscow and elsewhere continued. Negotiations to set up a dialogue between protest leaders and the Kremlin broke down on January 16, and by midwinter it appeared that Putin's strategy was familiar: to let time, small concessions and large handouts salve the social wounds.

Among the concessions, one at first appeared significant: in mid-January 2012 lame-duck president Medvedev presented Russian lawmakers with a draft proposal to restore the direct election of governors of Russia's regions. At first this seemed to be a direct and positive response to one of the demands of the protesters, but a reading of the proposal suggests that it would, if enacted, entail less democracy than met the eye. Political parties would be empowered to nominate candidates for election as governor, but not before there would be consultations with the president, "who shall set the procedures for such consultations." So much for democratic government; meanwhile, president-designate Putin took to denouncing "foreign elements" who allegedly were fueling instability within Russia and beyond. Leaving little doubt as to which foreign elements he had in mind, he identified "states that are trying to export democracy with the assistance of forceful, military means."

PUTINISTAN IN PROSPECT

In early 2012 it appeared certain that President Putin and his United Russia party would weather the electoral storm and assume the powers of the presidency in May 2012. In the United States, President Obama's policymakers and their advisers were preoccupied with East Asian strategy and Arab-world issues, seen as the salient concerns of a future in which an American military constrained by budget cuts would be required to limit its aspirations. The domestic upheaval in post-election Russia temporarily put the American-Russian quarrel over missile-defense systems on the back burner, and the larger issue of Russia's future in the new Putin era seemed to be below America's radar. But in speeches and writings, Putin had been making clear his plans for a country he deemed to have rescued from disintegration in the post-Yeltsin years.

In this vision, Putin plans to transform the political and economic geography of Eurasia from Kaliningrad to Khabarovsk and from the White Sea to the Caspian. Enabled by petrorubles and facilitated by state ineptitude throughout much of the Russian periphery, Putin's project will not repeat the fiasco that was the post-Soviet Commonwealth of Independent States (CIS). As Russia's ruler, Putin will have the power, the resources, and the blueprint for the resurrection in a new guise of the empire whose demise he publicly and frequently laments.

The evidence has been accumulating in various forms. Russian troops began arriving in Kyrgyzstan in 2002, shortly after Putin's first inauguration and ostensibly to counter Islamic terrorism—but more pertinently to offset the American military presence there. Although Russia's armed intervention in, and dismemberment of, neighboring Georgia occurred in 2008 on President Medvedev's interim watch, it would not have happened without Putin's authorization and sent shock waves through Russia's "Near Abroad." In late 2011, Moscow made the crucial U.S. military base at Manas in Kyrgyzstan an election issue there, urging its closure when its role in NATO's Afghanistan campaign remained crucial.

In the economic arena, Putin's design envisages a new Common Economic Space (CES), initially a customs union that links Russia, Belarus, and Kazakhstan and touted as the forerunner of a wider alliance that will move toward deep economic integration including the adoption of a common currency envisioned as a future competitor for the troubled euro. Formally inaugurated on January 1, 2012 with the accession of Belarus, the CES explains Putin's long-term lukewarm posture toward Russia's WTO candidacy, an apparent crack in the Putin-Medvedev alliance as the latter has actively pursued Russian membership.

Putin's pursuit of a Greater Russia rising among world powers seeks to extend Moscow's reach and expand on the formula that brought Russia back from the brink: centralized power and authoritarian government, military intervention within and outside the state when necessary, the exploitation of energy resources not only for state revenues but also for coercive purposes (as in the case of Belarus), and the use of Russia's nuclear arsenal and missile capacity as factors in strategic negotiations. The rewards of a wider CES will range from the reincorporation of Russian minorities (notably in northern Kazakhstan) and the reintegration of major energy reserves such as those in Turkmenistan to the compliance of governments and regimes in the Near Abroad still carrying the earmarks of their Soviet predecessors (Uzbekistan). A future Putinistan does not bode well for representative government in Central Asia.

But Putin's grand design hinges on one crucial component: fractured and fractious Ukraine. With its large and geographically clustered Russian minority, its Black Sea frontage, its oil and gas pipeline corridors, its industrial and agricultural output and potentials, and with Russia as its dominant trading partner, Ukraine is as important to any future Moscow-centered CES as it was to the communist empire of the past. Russia's heavy-handed interference in Ukraine's chaotic democracy and Moscow's cutoff of natural gas supplies to (and through) Ukraine during a price dispute in the 2008-2009 winter might suggest that Ukraine's future should lie with the European Union, not with Russia, but with the installation in 2010 of pro-Russian president Victor Yanukovych the tide seemed to have turned in Putin's favor. Soon after Yanukovych's election, Ukraine's government agreed to extend Russia's lease on the Black Sea naval base at Sevastopol for 25 years, for which Ukraine received a discount on the price of Russian natural gas. And in Ukraine's capital Kyyiv (Kiev), habits of old soon resurfaced: the imposition of restrictions on the media, the reversal of democratic reforms instituted by the previous administration, the termination of public debate over the Great Famine of 1932–33 as Soviet genocide. With Europe in disarray and Putin back in charge in Moscow, it is not difficult to envisage a new map of Eurasia—and a growing role for Russia in the world.

11

— — — —

AFRICA IN A GLOBALIZING WORLD

INFORMAL BUT PERSUASIVE SURVEYS INDICATE THAT, AMONG GEO-graphic realms competing for American attention, Africa ranks dead last—the "real" Africa, that is, the Africa lying south of the vast Sahara, the Africa of the defiant Ashante and the mysterious builders of Zimbabwe, of the powerful Zulu Empire and the stone city of Zanzibar, of the bustling markets of Dakar, the oil platforms of the Niger Delta, and the gold mines of the Witwatersrand. It is the Africa of the vast Congo Basin and the great Kilimanjaro, of Victoria Falls and Table Mountain, of the Great Lakes and the Kalahari. There is nothing like it in the world, its diverse peoples a kaleidoscope of cultures, its endangered primates a mirror of humanity, its dwindling wildlife populations a fading link with the early Tertiary.

It is also *terra incognita* to Americans more than any other part of the world. When Africa does gain America's attention, it tends to happen because of civil wars, health crises, natural disasters, or terrorist attacks, and only rarely because of the kinds of positive developments that occasionally emerge from other parts of the world. When the murderous dictator Abacha ruled Nigeria, he and his excesses were regular fodder for United States newspapers, but when the country achieved a remarkable, generally peaceful transition to democracy and President Obasanjo was elected, his tribulations in Africa's most populous and religiously divided country gained far less attention. South Africa's dramatic transition from apartheid to democracy generated a brief surge of interest and endowed President Mandela with celebrity status, but how much attention are the United States media giving today to what is by many measures Africa's most important country?

United States leaders do from time to time signal momentary awareness of Africa's plight and use it to make high-profile forays to states deemed deserving, as did President Obama in 2009 when he visited Ghana to recognize its importance as an emerging democracy, or highly publicized commitments to help solve Africa's problems, as did President George W. Bush during his first term with a $15 billion fund to combat AIDS (but also to promote his constituency's views on family planning). When Bush's first secretary of state, Colin Powell, was appointed he raised the hopes of many when he stated that Africa would rank among his highest priorities during his term in office. The events of September 11, 2001, and their aftermath ended that initiative before it could begin.

In the public eye, therefore, Africa south of the Sahara, the ancestral home of hominins and the cradle of humanity, the setting of our first communities and the scene of our first cultures, the place where we made our first tools and spoke our first words, the theater of our first artistic expressions and the landscape whose Pleistocene variability would propel our ancestors into Eurasia and the world—that Africa is largely forgotten. Originally we are all Africans, and to know ourselves better we should reconnect with the source. Our territorial and environmental imperatives began as African experiences, and some modern people, upon seeing Africa for the first time, experience an epiphany that may have scientific implications. The biologist Edward O. Wilson reports asking his students (and others) to draw what they conceive of as their ideal natural landscape; when he distilled a consensus out of thousands of such drawings that consensus resembled the East African savanna (Wilson, 1995). We may have left Africa, but Africa has not left us.

It is therefore appropriate to end our geographic journey in Africa, because this is where the human saga began and where many, indeed most, of the world's problems converge today. President Vladimir Putin of Russia and former president Thabo Mbeki of South Africa have this in common: they both recognize the effect of extreme poverty on human behavior. Putin, following a visit to the terrorist-targeted school in Beslan, railed against the "excesses" of democracy but also observed that people who are so poor that they have nothing to lose cannot be expected to accept their fate indefinitely. Mbeki attributed South Africa's AIDS epidemic to poverty rather than the virus that causes it; he was medically

wrong (at great cost to his nation) but socially right. If the world is becoming a global village, Africa is its poorest neighborhood, and when a
town's poorest neighborhood gets help, the whole community benefits.
If the European Union can siphon funds from its wealthiest members to
assist its less prosperous ones, then surely the world can find ways to
help Africa's neediest peoples.

African countries, unfortunately, rank high on the corruption ladder—
none is a Finland or a New Zealand, or even a Chile. So the argument
will be made that funds allocated are dollars wasted. But, as we will see,
there are other ways to help Africa. And there is ample proof all over
Africa that, given a chance, farmers and furniture makers, diamond cut-
ters and doctors are as good as any in the world. It is just that they need
a chance to prove their capacities in a world that has, in more ways than
one, turned its back on them and their region. In late 2004 the economist
Jeffrey Sachs tried to rally worldwide public support for a massive aid
program for Subsaharan Africa through a series of public lectures as well
as articles in the *Economist* and other journals, proposing a global "Marshall Plan for Africa" that would begin to address the most urgent of
Africa's ills. He proposed ways to target such aid, involving tens of billions of dollars contributed, like the European Union's subsidy funds, by
the richest nations of the world, directly to the neediest sectors of African economies. It was time, he argued, to invest in a better world by investing in Africa. The expense would be less than the annual cost of the
war in Iraq.

Africa suffers from a series of misfortunes, several of them geographic,
so severe that they have, in combination and in perpetuity, put Africans
at an incalculable disadvantage. This is not over—AIDS is only the latest
manifestation of just one of them—and it has reached the point where
what benefits much of the rest of the world can actually hurt Africa's
chances to catch up. Shortly after the remarkable transition to democratic government in South Africa, I happened to be in Australia where
I heard President Mandela's Afrikaner partner in this remarkable process,
F. W. de Klerk, present an address, unreported in the American media, in
which he argued that what the world needed was an Indian Ocean version of the Pacific Rim phenomenon: an Indian Ocean Rim anchored by
South Africa, Australia, Thailand, and western India that would transform the economic geography of the Southern Hemisphere and "carry
Africa into the twenty-first century." But in all of Africa, only South Africa and possibly southern Moçambique might see some benefit from

that vision. Australia and Southeast Asia were already in the Japanese-Chinese economic sphere. Unlike their putative partners, African countries tend to have little economic clout or political influence in the wider world. It will take more than free-market economics (not free for most African producers, anyway) to overcome the cumulative setbacks chronicled here.

EIGHT FORMATIVE DISASTERS

Africa's condition is a matter of statistical record. By any combination of criteria, from income levels to diets, infant mortality to life expectancy, health to literacy, Subsaharan Africa is the world's neediest geographic realm. When the Holocene epoch opened, Africa had for tens of thousands of years witnessed the trials and achievements of emerging humanity. But today, Africa and Africans, most of them descended from the first humans to walk on this Earth, suffer the effects of a combination of long-term disasters, beginning thousands of years ago, unlike those befalling any other corner of the globe. If non-Africans avert their eyes from the realm, it is because they see no hope and fear that the future will be still worse, not better. Those who could make a difference, including the so-called international community, meaning the powerful and richer states, leave Africa mired in its misery, sometimes blaming the victims for their problems. And it is not a matter of minimal subsidies; it is a matter of giving Africans a fair shake on world markets for products ranging from cacao to cotton. Champions of free trade deny African farmers opportunities on world markets through subsidies that, if ended, would send hundreds of billions of dollars to African cultivators.

But Africa is as powerless in the new unipolar world order as it was during the old bipolar world order, and it is a condition that has arisen from an incomparable series of misfortunes over the past 10,000 years, a combination of disasters that distinguished Subsaharan Africa from all other world geographic realms. Some other realms have suffered individual calamities that may exceed what happened to and in Africa, but none has experienced the coalescence of catastrophes that put Africa at the disadvantage it confronts today, and is reflected in indices ranging from life expectancies to incomes, food availability to infant mortality, and from disease incidence to education. Here are several sources of Africa's plight:

Climate Change

When the planet emerged from the extreme cold of the late Wisconsinan Glaciation starting about 18,000 years ago and warmed to the beginning of the Holocene epoch after the brief reversal to bitter cold during the Younger Dryas starting 12,000 years ago, the climatic and vegetative zones that lay compressed between periglacial and tropical latitudes began to shift poleward. That shift had momentous consequences for Africa, because the freezing up of Europe had pushed moist and moderate conditions into what is today the Sahara, where forests stood and streams flowed. Equatorial Africa was significantly cooler and equatorial rainforests had yielded to savannas, but now the warming brought rainforest expansion. As time went on and Egyptian civilization arose in the lower Nile River basin, the Sahara region became increasingly desiccated by 5,000 BP, creating a vast natural barrier between Mediterranean and tropical Africa. The Nile route continued to link Egyptian and African peoples, but the exchange of innovations between north and south was inhibited by environmental transformation across the entire northern bulk of the continent with consequences that would permanently isolate "Subsaharan" Africa. Eventually Egypt and Ethiopia, Morocco and Ghana would be worlds apart, separated by thousands of miles of inhospitable rock and sand, linked only by the most tenuous overland (in the west) and overwater (in the east) connections. Imagine instead an adequately watered, peopled, productive Africa extending to the very shores of the Mediterranean across from Europe, a Pan-African highway and road system linking Cairo to Cape Town, Algiers to Accra, Casablanca to Kinshasa.

The formidable Sahara barrier, its Sahel margin pulsating cyclically into what remains of its fragile adjoining ecologies, ensured that Africa would be regionally fractured and vulnerable to demographic crises, and that Africa would be "North" and "Subsaharan."

Ecological Impact

Not only did the Holocene's warming shift biomes from lower to higher latitudes; it also intensified equatorial climates in low-latitude and low-altitude Africa within and well beyond the margins of the Congo basin. That intensification produced higher temperatures, higher humidity, luxuriant rainforest growth and expansion, exuberant animal life from forest to savanna, the formation and filling of lakes and swamps, and, as it turns out, the final natural enlargement of the habitat of the great apes that had survived numerous previous environmental swings of the Pleistocene.

For Subsaharan Africa's human population, this global-warming episode had dire consequences. *Homo sapiens* had begun to emigrate from Africa via the southern end of the Red Sea as long as 90,000 to 85,000 years ago (an earlier attempt across the Sinai Peninsula had failed because of the sudden onset of the Wisconsinan Glaciation) but those who remained on the continent faced an environmental challenge that was to put them at a critical disadvantage when plant and animal domestication began elsewhere (Oppenheimer, 2003). While there is little doubt that African farmers did domesticate grains in West Africa as well as the Horn, animal domestication was another matter. Proliferating African wildlife was a threat to human existence, not an opportunity. While oxen, goats, and other animals were being domesticated in the Middle East and elsewhere in Eurasia, Africa's fauna defied domestication. As Jared Diamond reports, the only animal known to have been suitable for domestication in Subsaharan Africa was the guinea fowl (Diamond, 1997).

The heat and humidity of Africa's equatorial lowlands generated still another threat to African well-being: a host of diseases ranging from malaria to bilharzia and from yellow fever to sleeping sickness. Throughout the Holocene, the incidence of diseases rose, and even today the great majority of the approximately 1 million annual new victims of malaria are Africans. The equatorial environment nurtures countless vectors, from mosquitoes (malaria, yellow fever) and flies (sleeping sickness, river blindness) to worms and snails (the latter transmitting bilharzia), and people living in close proximity to wild animals, as well as consuming them, run further risk of transmission. Africa's problems in this regard are not just those of the distant past: sleeping sickness appears to have originated in West Africa as recently as the fourteenth century, and AIDS, Africa's latest scourge, began in the second half of the twentieth century. In 2011 Subsaharan Africa had 70 percent of the world's recorded AIDS cases and 75 percent of its deaths. Life expectancies had fallen dramatically throughout the realm and early projections indicated that AIDS would leave 20 million African orphans by 2010 (Altman, 2002). But in longterm perspective, AIDS is only the latest environment-related health calamity to strike Africa. The health of the general population has been poorer in Africa than anywhere else in the world for countless generations; remedies against African diseases have not been sought nearly as aggressively as has been the case for other areas of the world; and Africans tend to be too poor to be able to afford those medicines that do exist. For Subsaharan Africa's peoples, that combination of risk and neglect has been disastrous (Best and de Blij, 1977; Aryeetey-Attoh, 1997).

Divisive Islam

Surely the arrival of a great universal religion cannot be deemed disastrous for any realm or region of the world? In Africa's case, the Muslim faith came to unite, to sweep away local belief systems, to focus Africans' gaze on Mecca—but by its partial, incomplete penetration it also had the effect of dividing Africa along the Islamic Front referred to in Chapter 8. That line, in effect a narrow transition zone, extends from Guinea in West Africa to Ethiopia and the Somalia-Kenya border in East Africa. It cuts across the middle of Ivory Coast, divides Nigeria nearly in half, fragments Chad and Sudan, and cuts across the Ogaden in Ethiopia. Its social and political implications are momentous; it has precipitated wars costing millions of lives between Arab and African and between Arabized (Islamized) African and Christian and animist African.

Islam arrived in West Africa by caravan and in East Africa by boat, and in West Africa its proselytizers converted the kings of the great and stable indigenous states that prospered from the trade between peoples of the coastal forests and those of the drying interior. With the successful Islamization of Andalusia in full swing, the Muslims turned their attention to Saharan Africa, and by the fourteenth century they had done in the Saharan region what they had accomplished in Arabia seven centuries earlier. West African rulers made Islam the state religion, and huge, rich pilgrimages made their way along the savanna corridor from the states flanking the Niger River to the holy centers across the Red Sea. These pilgrimages, involving tens of thousands of people every year, played a major role in changing the ethnic map of the region: many stayed behind, on the way to or from Mecca, implanting their cultures from Nigeria to Sudan.

But Islam's southward march was halted by African resistance and by European-Christian intervention, and the Islamic Front came to reinforce Africa's division between Northern and Subsaharan. Now it was not just a desert but also a faith that reinforced the continent's partition, a further barrier between two realms and, ultimately, a source of devastating conflict.

The Depopulating Slave Trade

Putting people in bondage and working them to death was a worldwide practice (the word "slave" comes from the ethnic term *Slav*, Eastern Europeans enslaved by Muslim Turks). The ancient Greeks and Romans had slaves and the powerful enslaved the weak or the defeated in many, though not all, cultures. But nothing in the world compared to what happened to Africa when Europeans arrived not just to enslave Africans, but to capture, transport, and sell them overseas. True, Arabs had been carrying on a slave trade from East Africa for centuries before the Europeans organized their version of it, mainly from West Africa but also from other areas. It was the dimension, the sheer volume of the European slave trade that dwarfed all other forms of the practice (Curtin, 1969).

Much has been written about the numbers, New World destinations, and dreadful treatment of those who were taken in bondage from their African abodes and who survived the unimaginable Transatlantic voyage. Estimates of the dimensions of the traffic between 1700 and 1810 range from 12 million to more than double that figure; the world will never know (Eltis and Richardson, 2010). Still less is known about the consequences of the slave raids in Africa itself. Subsaharan Africa's population at the beginning of the eighteenth century may not have been much more than 90 million (the estimate for the global population in 1700 is 650 million), so that if the number of forced migrants was 15 million, this means that one-sixth of Africa's population was taken to the Americas (Fig. 11-1).

Anthropologists reckon that the impact of this terrible, catastrophic event can still be read in Africa's cultural landscapes. Not only were entire areas depopulated by the slave raiders as well as through the fighting that attended the campaign, but the price put on human heads set African against African and rekindled ethnic animosities. Children were orphaned or abandoned by the hundreds of thousands, crops lay unharvested, villages stood deserted if they were not burned. West Africa's forest-to-desert trade collapsed, the Islamized interior states broke up, and everywhere the social order disintegrated. Columns of chained Af-

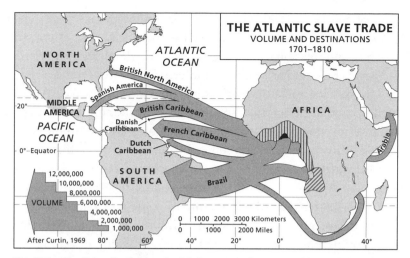

Fig. 11-1. The Transatlantic slave trade's impact can be measured in numbers, but never in terms of its lasting demographic and cultural impact on Subsaharan Africa.

ricans marched from ever-deeper inland "sources" to the coastal embarkation stations, never to see their homeland again.

By itself, a disaster of such magnitude would have lasting consequences for any cultural realm, but in Africa the slave trade was only one of a series. When the slave trade ended and the Europeans, now taking the moral high ground, closed the last of the Arab slave markets in East Africa, the damage done to Africa's social fabric was incalculable. But the foreign powers that had conducted it were still there, pursuing other goals.

Colonialism

Europeans had been testing African waters long before the slave trade started, and they would remain as occupiers long after it ended. European trading posts were founded along the West African coast in the 1500s to outflank the commerce that was still linking coastal states, savanna empires, and Saharan caravans, including the sizeable and lucrative trade in gold. Portuguese vessels rounded the Cape of Good Hope and entered the Indian Ocean, the Dutch established a supply station at Cape Town in support of their East Indies traffic, and the British and French, followed later by the Belgians, Germans, and Italians, staked their claims to African empire.

Most of this activity, though, was concentrated in Africa's coastal areas. Well into the nineteenth century the interior remained the sphere of explorers and missionaries, raiders and occasional traders. When territorial issues began to arise and the fateful Berlin Conference was convened late in 1884 to settle conflicting claims and to carve Africa up among the competing colonial powers, much of the continent's vast and remote interior was barely known to the contestants. Boundaries were drawn to codify spheres of influence, but in many areas there was not enough information to gauge how these colonial borders would affect local peoples. As a result, some of those boundaries split unified ethnic groups into different jurisdictions; others combined peoples with historic animosities. Still others disastrously interfered with seasonal migrations, depriving pastoralists and their herds of water and forage when Berlin's borders were imposed on the ground (Newman, 1995).

Colonialization meant exploitation, and exploitation required terror. In the current era of terrorism it is well to remember that a century ago African peoples were terrorized by European colonists on a "civilizing mission" that subjugated an entire culture realm, a mission that was carried out by a small minority of invaders whose principal means of maintaining control was fear. The Europeans' shared objective was the exploitation of Subsaharan Africa's natural resources, and in this pursuit they established administrative headquarters, transport routes, and ports, creating the beginnings of Africa's modern, disconnected infrastructure. To be sure, their methods differed: in general, British colonial practice was more benign than that of the French or the Portuguese, and the Germans and the Belgians were the most ruthless. The short-lived German occupation of four African territories was referred to in the European press as "colonization by the Mauser" because so many tens of thousands of Africans were killed by this efficient gun; the story of the Belgian King Leopold's "Congo Free State" ranks among the century's worst human calamities.

The world should reflect on what was done to the approximately 20 million inhabitants of the territory awarded to King Leopold during the Berlin Conference, because the aftermath still lingers in this blighted country: "King Leopold II . . . embarked on a campaign of ruthless exploitation. His enforcers mobilized almost the entire Congolese population to gather rubber, kill elephants for their ivory, and build public works to improve export routes. For failing to meet production quotas, entire communities were massacred. Killing and maiming became routine in a colony in which horror was the only common denominator. After the impact of the slave trade, King Leopold's reign of terror was Africa's worst demographic disaster. By the time it ended, after a growing outcry around the world, as many as 10 million Congolese had been murdered" (de Blij, 2004).

The destructive impact of imperial pursuits in Africa are well enough known today not to require further elaboration, but colonialism had a more subtle yet equally enduring effect. Subsaharan Africa in colonial times was culturally a highly diverse geographic realm, as the famous "Murdock Map" of ethnic areas, published nearly a half century ago, underscores (Murdock, 1959). But those lines delimiting the traditional domains of Africa's ethnic groups are deceiving. True, as in all parts of the world, there are well-defined ethnic borders in Africa. More often than not, however, the lines on the Murdock Map represent transitions, zones where African peoples with different cultural traditions lived together (Fig. 11-2). Already it was, in many areas, difficult to tell representatives

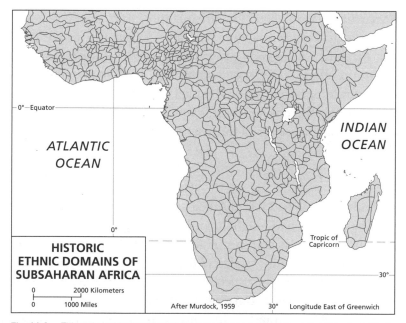

Fig. 11-2. This painstakingly constructed map known worldwide as the "Murdock Map" after its researcher and cartographer James Murdock, attempts to reconstruct what the ethnic mosaic of Africa looked like before the European colonial invasion began. As Murdock emphasized, the thin lines between ethnic domains often were transition zones, not sharp divides. And African views of land ownership and territorial boundaries differed markedly from those of the European colonists.

of one "tribe" from another. In present-day Rwanda and Burundi, for example, where sedentary Hutu farmers had been invaded by pastoral Tutsi, some Hutu had become landowners and herders in the Tutsi tradition and a process of cultural convergence was under way.

As John Reader recounts, one common practice among colonial powers with otherwise divergent policies was to classify all their African subjects on the basis of ethnicity. Whatever the objectives (to facilitate "indirect rule" in British dependencies, to organize forced cropping on Portuguese farms, to collect taxes everywhere), the policy had divisive effects that would long outlast the colonial era. It was retribalization that enhanced the status and power of some peoples while diminishing others; it assigned Africans with mixed backgrounds to often unfamiliar social status; and it laid the groundwork for conflict in postcolonial states (Reader, 1998). The policy reached its apogee in apartheid South Africa, where the variant was domestic, not European, colonialism and the results continue to plague the multiethnic nation.

It is tempting to speculate on the kind of Africa that would have emerged in the absence of colonial intervention and its associated boundary demarcation, land alienation, urbanization, and other consequences of what was in effect Subsaharan Africa's first experience with globalization. Africans had already proven their capacity for building long-lasting, multi-ethnic states; they had already built major cities and engaged in long-range, interregional trade (Newman, 1995). We will never know, but of this there is no doubt: the colonial episode was another in Africa's unparalleled series of formative disasters.

The Cold War

In the aftermath of World War II Africa's prospects appeared to improve. The colonial powers had fought among themselves and weakened their ability to hold overseas empires (France and the Netherlands in Southeast Asia, Britain in South Asia), and the superpower Soviet Union proclaimed itself a bulwark of anticolonialism. By the 1950s, African resistance had grown strong and negotiations were under way to facilitate the transfer of power from London and Paris to such capitals as Accra, Lagos, and Dakar. Rebellions and larger conflicts pitted Africans against Europeans in such territories as Kenya, Congo, and Angola, and white settlers held back the decolonization process in (then) Rhodesia, but within little more than two decades, starting with the Gold Coast (re-named Ghana) in 1957, Africa's colonial era ended and several dozen newly independent states appeared on the map.

Inevitably, such rapid transition from colony to statehood produced crises and conflicts. Some colonial powers had done more than others in fields such as higher education and civil administration to prepare their dependencies for sovereignty. Britain left Ghana with a Westminster-model government and a strong economy and helped Nigeria deal with its regional schisms by developing a federal system. Numerous African university graduates from African as well as British universities gave these new administrations authority as well as authenticity. The Belgians, on the other hand, had done little to generate a college-educated cadre of Congolese although they knew that independence was inevitable, and the result was chaos. In addition, conflict quickly arose among nations in states that had held comparative power and privilege during the colonial period and other nations that saw, in independence, an opportunity to improve their position. Kenya's Kikuyu, Uganda's BaGanda, Nigeria's Yoruba, and many other prominent African nations were thus challenged by others within the state, so that politics soon tended to take on a "tribal" tone.

Such transition-engendered conflicts arose at the worst possible time for Africa, because many of them were magnified by the superpowers competing in the Cold War. The Soviet Union, itself a colonial empire, saw an opportunity to legitimize its anticolonial credentials by supporting those African groups that professed leftist leanings. The United States, which had vigorously supported African independence, had appointed a secretary of state for African affairs, and sent thousands of Peace Corps workers to Subsaharan Africa, nevertheless acted also to obstruct Soviet designs. The CIA was implicated in the murder of the first Congolese prime minister. From Ethiopia to Angola, proxy wars raged and hundreds of thousands of ordinary Africans died of war, starvation, and dislocation.

The Cold War caused the United States to turn a blind eye to the excesses of African rulers who could be trusted to oppose communism, including the Congo's Mobutu Sese Seko, Ethiopia's Haile Selassie, and others; along with the former colonial powers it tolerated murderous tyrants like Idi Amin (Uganda) and Jean-Bedel Bokassa (Central African "Empire"). With the Soviet Union encouraging the likes of Mengistu

Haile Mariam (Ethiopia) and Agostinho Neto (Angola), millions of Africans had traded foreign imperial rulers for domestic ones under the sway of foreign powers. It was another damaging turn of events for a realm falling ever farther behind the rest of the world.

Globalization

Once again the world is changing dramatically, as it did during the era of colonialism and during the wars of the twentieth century. Today the changes come through a series of interrelated processes referred to as globalization, the breaking down of barriers to international trade, the stimulation of commerce, the revival of long-dormant economies, the stimulation of social, cultural, political, and other kinds of exchanges. In the rich and powerful countries of the world, globalization has enthusiastic support although small but vocal minorities oppose it. But in the less privileged parts of the world, where participation in the process is not easily achieved, globalization is seen by many as a threat, not as an opportunity.

Analyses of the worldwide impacts of globalization underscore the negative effects the process has on many countries in Subsaharan Africa. While it is evident that an openness to globalization in its various forms is advantageous to countries already comparatively wealthy or on a relatively rapid developmental path, the poorest countries, most of them in Africa, with weak economies propped up by unwise tariffs and other trade barriers, lag ever farther behind. Thus globalization is increasing, not decreasing, the gap between poor and well-off; statistical averages notwithstanding, the lot of the average African has not improved in the era of globalization. The privatization of some of the few state-held assets in these poorest countries, or the construction by a global corporation of a coastal casino, luxury hotel, or beach resort in Moçambique or Tanzania may suggest modernization and may be reflected in "development" data, but does nothing to change the fundamental condition of the state. The geographer Robert Stock has described the poverty-stricken condition of Subsaharan African cities, many of which are in worse condition today than they were at the end of the colonial era (Stock, 1995). Urban landscapes reflect Africa's remoteness from the forces of change that are transforming urban centers from Santiago to Seoul.

Is globalization a disaster for Africa on a par with the other seven chronicled here? It is yet another process of extra-African origins that has a deep and long-term impact on Africa's internal condition and external position. Globalization's meager penetrations in Africa (South Africa excepted) have stoked corruption and threatened fragile cultures and ecologies. South Africa, open to globalization's impacts after the end of the apartheid era, came to rank as the world's most unequal society within just one decade. Analysts sometimes observe that Africans have themselves to blame for the loss of opportunity associated with globalization's slow march in this realm, but globalization is a power play, and power on the global stage is not an African asset. The cumulative consequences for Africa constitute a terrible misfortune.

The Failure of Leadership

During the 1960s, when Europe's African dependencies followed Ghana's lead and marched toward independence, African leaders garnered worldwide admiration. Many of these leaders had spearheaded independence movements long before the colonial powers began to contemplate a transfer of power, some led armed insurrections and many served time in prison. The exhortations of Ghana's Kwame Nkrumah inspired African nationalists everywhere; the detention of Kenya's Jomo Kenyatta during the Mau Mau uprising symbolized the struggle for all. When France offered a strings-attached sovereignty to its West and Equatorial African colonies and Guinea was the only prospective state to turn this down, Sekou Touré became a hero from Dakar to Durban. In the early years following independence, these "fathers of the nations" appeared to prove eloquently what Africans were capable of, and what the colonists had mindlessly suppressed. Tanzania's Julius Nyerere translated Shakespeare into Swahili; Senegal's Leopold Senghor was elected to the French National Academy of Sciences. African leaders spoke eloquently at United Nations conferences and meetings of nonaligned states during the Cold War.

But the route from father of the nation to usurper of the people turned out to be a short one. Military coups ended many brief experiments in democracy and ousted elected leaders, and elsewhere the legitimate successors to the nations' founders had fewer principles than their predecessors had. Even as murderous regimes such as those of Amin (Uganda), Bokassa (Central African "Empire"), Abacha (Nigeria), Mobutu (Congo), and Doe (Liberia) shocked the world with their excesses, other, more stable African states began to appear on lists of the world's most corrupt countries. Kenya, where Daniel arap Moi had succeeded Jomo Kenyatta, headed this depressing roster.

In Southern Africa, the long-imprisoned Nelson Mandela made possible that almost unimaginable, peaceful transition from apartheid to democracy, setting one glorious and final example of African leadership and statecraft in the liberation era. But in South Africa's neighbor, Zimbabwe, the revolutionary hero Robert Mugabe turned into another of Africa's destructive tyrants—in the process ruining one of the continent's most promising economies. When Nelson Mandela was democratically succeeded in South Africa by Thabo Mbeki, the new South African president found it difficult to express toward Zimbabwe the moral standards he inherited from the founder. Mbeki's successor, Jacob Zuma, showed even more weakness.

African leadership has too often failed in Africa, but the leaders of the "international community" have failed Africa as well. Former European colonizers turned a blind eye to the misdeeds of the dictators with whose countries they did business; Cold War politics prevailed over principle when ruthless rulers took sides; and more recently the convulsions of the Congo and its eastern neighbors, costing more than 5 million lives, failed to stir the world into the massive intervention the region, by some measures, was surely owed.

Today there are some brighter prospects: a return to democratic government in Nigeria and flawed, violent, but essentially democratic elections in Kenya, peace and reconstruction in Angola, stability and progress in Senegal, Ghana, and Zambia, and other signals of a new era. But a half century of widespread failure has taken a toll that will long be imprinted on the African map.

CHINA IN AFRICA

Geographic imprints have a way of persisting and enduring. In an earlier chapter we noted that for all its economic transformation and dramatic modernization, China still exhibits the hallmark of its demographic history, the concentration of its population in four river basins in the eastern third of the country. For all its effort to unify and integrate, Europe still is dominated by the powerful states that have long ruled a smaller and less influential roost. So it is in Africa. The influences of the events and circumstances chronicled in this chapter still mark the African map, from the surge of Islam in the north to the imposition of its boundary framework by the European colonialists, and from the Transatlantic and Arab slave trade to the exploitation of its natural resources.

Today a new era is in progress. Globalization takes many forms under many guises, from neocolonial to neoliberal, and if, as many observers report, Subsaharan Africa has been left behind in certain arenas, it has a new trading partner making up for it: China. The Chinese presence in Africa is ubiquitous and pervasive. Its goals range from the purchase (and control) of raw materials to the implantation of cultural icons; its investments range from mines to farmland; its objectives are diverse but do not include interference in local politics. While it is much too early to tell whether China's growing presence in Africa will ultimately accrue more positive results than negative ones, there can be no doubt that a new era has arrived. Subsaharan Africa has commodities China needs, notably in the energy field but also in metals and minerals, and China has the money to invest not only in exploitation but also in infrastructure and related projects. From ports to railroads and from bridges to real estate, China is changing the economic geography of Africa.

China is not, as some fearful analyses have suggested, trying to become Africa's new colonial master. But make no mistake: as elsewhere in the world, Chinese companies, state as well as private, have more than commercial objectives: invariably they want to proclaim, in word and deed, China's rising status as a world power able to accomplish things the "old" powers cannot. They want to buy commodities, but they also want to create African markets for Chinese products, encourage settlement in Africa by Chinese migrants, proclaim the virtues of Chinese culture and Chinese development models (in contrast to Western ones), and gain support from African governments and regimes in the international arena on issues ranging from Taiwan to votes at the United Nations (like Russia, China in 2011 voted against imposing sanctions on Syria's regime even after more than 3,000 of Syria's citizens had been killed by "security" forces there during the "Arab Spring"). To make it clear: China is not in the business of democratization or regime change. It will do business with whoever is in control, and that includes democratically elected presidents as well as the likes of Zimbabwe's Robert Mugabe and Equatorial Guinea's Brigadier General Teodoro Obiang.

This has consequences. Labor relations in Africa involving Chinese companies have at times been difficult, even violent. The so-called "Copperbelt" in Zambia has been the scene of accidents and retaliation: in April 2005, for example, a blast destroyed a Chinese-owned explosives factory in Zambia, killing more than 50 local workers; allegations of worker mistreatment by Chinese foremen had already led to kidnappings and strikes. When the local media reported that a Chinese-owned mine was forcing laborers to go down the elevator shaft to work in their jeans and shirts, without any safety equipment, another round of accusations and mistrust ensued, canceling a planned visit by China's president. The growing Chinese presence in Zambia also became a key issue in the country's presidential campaign, with one candidate running on an anti-Chinese platform—and the Chinese Ambassador threatening to cut diplomatic relations between China and Zambia if he won. This appeared to undercut Beijing's self-proclaimed noninterference rule and sent shudders across the continent.

The often-unfavorable working conditions on Chinese projects and associated low wages have led some African governments to try legal means to protect their workers, but in reality the workers' options are few and such rules are easily circumvented. Meanwhile China's clout continues to grow: the China Export-Import Bank is the largest supplier of loans to the African realm, and in its quest for agricultural land China reportedly has some 10,000 farm "experts" in Africa (Carmody, 2011). China is funding the rehabilitation of Zambia's rail outlet via Angola, the Benguela Railway to the (also revived) port of Lobito. There are reports of a proposed $6 billion project for a Chinese-built and operated port near Lamu on Kenya's north coast, which might someday export oil from the newly independent state of South Sudan. Other rumors have China building a military base at Kamina in the mineral-rich southern Congolese province of Katanga, on the other side of the border from Zambia's Copperbelt.

Indeed, there is no African country today without a Chinese presence, and it is too early to discern which way the relationships between the rising superpower and the ultimate global periphery will go. As noted earlier, one objective for China is to create markets for Chinese products in Africa, which is already having major impacts on still-modest African manufacturing. Tanzanian shoemakers, for example, were pretty much put out of business by cheap Chinese imports. According to Padraig Carmody, cheaper Chinese doughnuts are putting local bakers out of business in Douala, the capital of Cameroon; in many places Chinese shopkeepers are resented by local traders.

But at the other end of the social ladder, things are different. Western governments try to discourage companies from doing business in African countries with authoritarian governments, and do not invite the leaders of such countries to meetings in their capitals. China, on the other hand, takes a more flexible approach, entertaining African elites no matter what their record on human-rights or democracy. The most recent meeting of Chinese leaders and African leaders and elites (2009) brought those of 41 African states to Beijing, reportedly the largest meeting of its kind ever held. Official Chinese visits to Africa strengthen ties thus established, with long-term objectives high on the agenda.

Given China's growing demand for what Africa has to offer and its ability to pay, Africa's own requirements in a globalizing world, and the changing balance of economic power in the world generally, it can be argued that China's penetration of Africa is the most portentous development to affect the continent since the end of the proxy conflicts of the Cold War. Whether it will turn out to be a triumph or another of Africa's disasters only time can tell.

WHY AFRICA MATTERS

"Always something new out of Africa," exclaimed a Roman emperor 2,000 years ago, and so it is today. In the rainforest of the northeastern Congo, workers paid a pittance for their labors are digging from the ground coltan, a raw material used in the manufacture of mobile phones. From uranium in the atomic age to oil in the fossil-fuel era, Africa has always had what it takes—for the rest of the world.

But concern for Africa's well-being should not focus on the relentless acquisition of its commodities. Africa's problems and the world's concerns coincide because the world is functionally shrinking, and when one of the neighborhoods of the "global village" suffers more than any other from a combination of maladies, the remedy benefits all. So assisting in the recovery of Subsaharan Africa is not mere altruism; it is a matter of self-interest for the rest of the world—and especially for the United States.

In the first place, social stability in Africa (as elsewhere) is a precondition for progress. Recent events in West and Equatorial Africa gave rise to fears, prominently expressed by Robert Kaplan, that the civil wars in Sierra Leone and Liberia would spread widely, precipitating a "regional anarchy" (Kaplan, 2000). Instead, a small contingent of British troops quelled the violence in Sierra Leone during the late 1990s and arrested the movement's leader, Ferdie Sankoh. And when American forces failed to intervene in Liberia an army of Nigerian peacekeepers drove the country's ruler, Charles Taylor, into exile in 2003 (he was held responsible for the deaths of more than 300,000 people by starting ethnic conflicts that spilled over from Liberia into Sierra Leone). In Equatorial Africa, what started as another in a series of Tutsi-Hutu conflicts in Rwanda spilled over into the eastern Congo and ended the rule of the feared Mobutu, but not before drawing in an assortment of regional profiteers who saw economic opportunity in the chaos.

The greatest danger to African stability lies along the "Islamic Front" referred to earlier (Fig. 7-2). Several postcolonial states are bisected by this cultural transition line with attendant risk, including Côte D'Ivoire (Ivory Coast), recently the scene of bitter conflict between southerners and northern Muslims, and Nigeria, where 12 of the federation's 36 States have adopted Islamic sharia laws (Fig. 11-3). The schism between north and south in Nigeria is fraught with danger, and cultural distinctions have grown sharper over the past decade as hundreds of thousands of southerners living in the north have moved away. The stability of Nigeria, cornerstone of Africa's historic cultural core, is indispensable to Africa's future, and the country's return to democracy should be acknowledged through substantial international assistance toward its infrastructure needs.

Islamic outposts as far away as Cape Town form a reminder of the susceptibility of Subsaharan Africa to Islamic proselytism, religious as well as political. The vulnerability of the poor, the frustrations of neocolonialism real and perceived, the association of Christian faiths with imperial histories, fears of globalization, and promise of empowerment give Islam's proponents opportunities in Africa that may change the ideological map of the world. To counter this prospect, Western involvement in Subsaharan Africa must grow exponentially, supporting democracy, building infrastructure, opening markets, and strengthening linkages.

Which leads us to the third consideration under the present rubric, the question of products and commodities. Western involvement in Africa should go far beyond the purchase of oil and minerals, and requires sacrifices from all wealthy countries, Western and non-Western. The matter of agricultural subsidies ranks high in this category. The majority of Africans today still make their living as farmers, subsistence as well as commercial, and for the latter, any unfair competition on "free" markets has devastating results. But that is exactly what African farmers face,

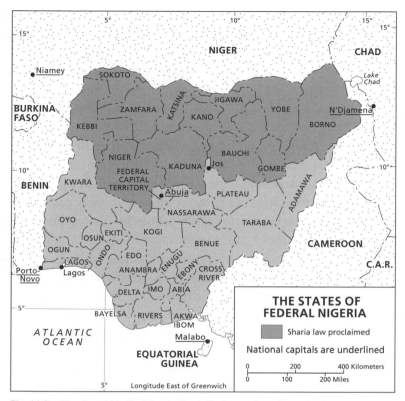

Fig. 11-3. Nigeria astride the Islamic Front. Twelve northern States in Nigeria's federation have adopted Sharia law; friction between Muslims and non-Muslims along the divisive cultural border continues to take a significant toll.

from France to Japan to the United States. African farmers earn very low per-hour wages, and they are able to market their tea, cotton, cacao, and bananas at very low prices. But they cannot compete with farmers who receive state subsidies to plant, harvest, export, and market their produce. It is up to the rich countries to phase out these subsidies—if they really believe their own rhetoric about free trade—and to give African farmers their chance. A recent World Bank estimate indicates that African farming would benefit to the tune of more than $200 billion per year, more than 20 times the financial aid currently given to Africa by donor countries.

African farming is in a poor state, and to be sure there is more involved here than the policies of the richer countries. African governments themselves have failed to support their farmers in several ways: by keeping prices of staples artificially low, pleasing urban residents but costing farmers dearly, by making it difficult for farmers to secure loans, by investing heavily in ill-conceived industrial projects rather than the important agricultural sector of the economy, and by denying women, who do much of the actual work in the fields, credit opportunities to start and maintain their own farms or businesses. Add to all this the severe seasonal fluctuations in Africa's weather and the marginal productivity of much of Africa's soil, and it is clear why Africa's per-capita farm production has been declining for more than two decades.

If there is little incentive in the richer world to give African farmers a break, there is ample reason to deal more fairly and cleanly with African governments when it comes to the continent's energy resources and minerals. World-market prices for commodities fluctuate, and declining earnings from metals that fueled exploitation by colonial powers have strained the economies of African states. Zambia, heavily reliant on copper, has seen its income shrink not only from reduced market prices but also because the landlocked country's most efficient outlet to the sea, via Angola, was destroyed by the prolonged conflict there. South Africa's economy has suffered from declining gold prices and increasing labor costs. But where one country ails, another thrives: Botswana's income from diamonds has soared, and a combination of stable, democratic government and judicious economic policy has moved Botswana into the middle-income category of African states, still a rare phenomenon.

The most significant development since decolonization, however, has been in energy resources and their exploitation. Africa is proving to contain far larger oil reserves than was estimated only 30 years ago, and today not only Nigeria (which supplies the United States with about 12 percent of its annual oil imports) and Angola but also Gabon, Chad, and Equatorial Guinea, Sudan, and South Sudan are important producers.

The implications of Africa's rise among oil-producing regions is obvious in this time of strategic uncertainty and potential disruption in the Middle East. West and Equatorial African reserves lie directly across the Atlantic from the world's thirstiest consumer without choke-point perils or pipelines as vulnerable as those elsewhere. Thus the creation and maintenance of stable political and trade relationships between the United States and Africa's producers is in America's self-interest.

The implications for the citizens of the producing countries, however, are troubling. Peoples whose historic homes are in the oil-rich Niger Delta, for example, have waged a decades-long struggle to assert their rights to a far larger share of the proceeds than they have been awarded

HERE WE GO AGAIN

Angola has become China's leading oil supplier, fueling an unprecedented boom in the capital, Luanda. Angolan businessmen are partnering with Chinese entrepreneurs in syndicates that, according to a briefing in *The Economist,* are doing secret deals that may not only be depriving some of the world's poorest people of desperately needed income but, elsewhere in Africa, may be perpetuating violent conflicts.

When Chinese private companies enter the African market, they take on obligations to win their mining licenses. These obligations include infrastructure projects such as low-cost housing, water-supply systems, roads and railways, and hydroelectric plants. All too often, the total investment in such projects is small compared to the profits generated by oil, a bad deal for African states. Or they don't happen at all. Six years after the syndicate started doing business in Angola, more than 90 percent of the residents of burgeoning Luanda remained without running water. Meanwhile, "oil contracts are treated as state secrets . . . accusations of personal enrichment percolate up toward the top of the state structure . . . in 2006 [it was alleged that] $2 billion of Chinese money intended for infrastructure projects had disappeared, [allegedly] transferred to private accounts in Hong Kong by senior officials."

"Rather than fixing Africa's lack of infrastructure," the briefing concludes, "Chinese entrepreneurs and Africa's governing elites look as if they are conspiring to use the development model as a pretext for plunder." Will the Chinese connection turn into Africa's Ninth Disaster?

—Quoted from "The Queensway Syndicate and the Africa Trade," *The Economist,* August 13, 2011

by rapacious regimes and voracious corporations. In 1995 the Abacha regime executed by hanging nine leaders of the Ogoni, a people in the vanguard of this movement, while a president-elect languished in jail. But neither trading-partner governments nor the oil corporations themselves risked their relations with Nigeria's dictatorial clique over this issue. To those Nigerians then old enough to remember, it was all too familiar.

Events in the Niger Delta, in war-torn Angola (where oil and diamonds fueled the conflict), and in Sudan (where northerners ruthlessly dispossessed southern villagers beneath whose soil oil was discovered before the South voted to secede) appear to have led to greater scrutiny of the actions of governments and oil companies.

Whatever the outcome of this initiative, there can be no doubt that the world needs Africa's commodities. And now a new trader is on Africa's block: China, ravenous for raw materials. Already, China's huge purchases have the Australian economy booming. But what Africa needs is what Australia already had: a thriving economy of its own. In this respect the consumers of Africa's products must do more to help the source.

Africa matters to the rest of the world in still another way: as the global geographic realm most severely afflicted by dispersible diseases. Again the massive medical intervention Subsaharan Africa needs is not a matter of altruism: it is in the entire world's interest to improve public health in this, the sickest part of the planet.

What once was Africa's problem—the warming episode that made its tropics a vast incubator of a host of diseases—is now a global problem (Gould, 1993). Africa's links to the outside world may be weaker than those of other realms, but jet aircraft, passenger and cargo ships, and other conveyances can carry African maladies in human and animal vectors around the globe in a matter of hours (the first case of AIDS in North America may have been transmitted by an infected individual who traveled by jet from Kinshasa to Brussels to Chicago). The world is fortunate that a disease far more infectious than AIDS has not emanated from Africa to kill millions in short order, but medical geographers predict that such a time will come unless a huge, focused effort is made to improve public well-being here at the heart of the Land Hemisphere.

My colleagues who study such matters point out that dangerous diseases in the richer world for many years have received far more attention than diseases long confined to Africa. If malaria had been a midlatitude malady, they argue, it would have been wiped out long ago. From polio to smallpox, many rich-world diseases have been conquered. When AIDS took on serious dimensions in the United States, the medical establishment took it on full force, and today it is possible to keep HIV in check, perhaps indefinitely. Had AIDS remained confined to tropical Africa, it is unlikely that such a campaign would have been waged.

We know more about the endemic and epidemic maladies of Africa today because research is delineating them and because international agencies are reporting their prevalence, and because the media now vividly describe events such as outbreaks of the feared ebola fever and other dramatic but local health crises. African epidemics rarely become pandemics, as AIDS did, so the perceived threat of African infectious diseases spreading beyond Africa is low. But the risk is real and growing, and the best way to interdict it is to raise the level of well-being of Africans everywhere. Already, the World Health Organization (WHO) under UN auspices inoculates millions of African children against a range of diseases, but many more do not get the protection they need. As a result, infant and child mortality rates in many African countries still rank among the highest in the world. Those who survive are likely to suffer from several of the endemic diseases of Africa against which protection is still inadequate.

Attacking Africa's diseases (much progress has been made, for example, against river blindness) is one thing; making medicines affordable is another. As the AIDS disaster showed, even when Western drug companies had achieved remedies that can keep the virus at bay, the cost, while declining, remained too high for most Africans to afford. When the average annual income of a population is well below $800— the approximate figure for Moçambique in 2011—the $300 monthly cost for the full range of needed drugs is simply out of the question. Even substantial per-capita subsidies will go only part way to making AIDS medication available where it is needed most.

Improving the public-health situation in Subsaharan Africa should be a huge, international project not only to benefit Africans but also to reduce the risks Africa's medical situation poses to the rest of the world. In addition, it is an indispensable element in any economic recovery Africa may make. The stakes are high and the potential returns immeasurable.

To the multicultural nation of the United States, Africa matters in still another way. The United States never held African colonies (its special relationship with Liberia arose from the repatriation of Africans who had come to America in bondage) but the African American population in this country is larger than the population of all but five Subsaharan African countries. In truth, Africa is America's second cultural source, after Europe, but the Americanization of the African minority has weakened the links between Africa and America's African diaspora. A recent press report suggested that more African Americans vacation in Europe than in Subsaharan Africa, and knowledge of Africa among African Americans appears to be no stronger than among whites.

This suggests a host of opportunities. In Brazil, which has a far stronger African imprint than the United States, interest in Africa is intense to the point that a major two-way flow of visitors has developed between the country of Benin (formerly Dahomey) in West Africa and the State of Bahia (centered on Salvador), following the discovery that many Africans were taken directly from Benin to Bahia during the period of slavery. The survival of African cultural traits in Bahia and the identification of ancestral homes in Benin created a remarkable Transatlantic cultural revival. For African Americans in the United States, such linkages may not be as discernable, but surely the recovery and well-being of Africa is a prospect of intense interest to all, and African American commitment and involvement would be key ingredients in the process. In this context, it is not difficult to hear the echo of former secretary of state Powell's inspiring words.

Africa today is on the move. While the sale of raw materials currently drives many national economies, it is noteworthy that four of the world's fastest-growing economies in 2009 were African: Angola, Ethiopia, Rwanda, and Equatorial Guinea (two of the four without oil exports) and that in 2010 a dozen Subsaharan-African countries ranked in the World Bank's Upper and Lower Middle Income categories. Indeed, the per-capita gross national income (GNI) of seven African countries in 2010 was higher than that of their trading-partner, fast-growing China.

And Subsaharan Africa, still the least-urbanized of the world's dozen geographic realms, is fast catching up. Megacities such as Lagos and Kinshasa are burgeoning, smaller cities and towns are growing, and in those cities and towns millions of arrivals drive a huge informal economy. From Dakar to Durban, these are globalization's contact points marked by financial districts and gated communities, shopping malls and service sectors. The capitals are more than governmental headquarters: here the battle is waged between autocrats and progressives, conservatives and modernizers. Paul Collier has drawn attention to the cadres of courageous, mostly young, African intellectuals who risk their careers, sometimes their well-being, in the fight Africa must wage: the battle against corruption (Collier, 2007).

African countries rank dismally low on "transparency" charts, but they are not alone. Nigeria, at 2.4 in 2010 (0.1 is worst, 10.0 unattainable best), ranked above Russia (2.1) and on a par with Ukraine. African rulers and leaders have stashed away countless billions in foreign bank accounts, thefts that have much to do with the "corruption index" several agencies publish annually. But it is "petty" corruption that afflicts ordinary people in everyday life, whether it occurs when opening a bank account, getting a driver's license or even just traveling a road past a "check point." Living with corruption of this kind saps confidence and kills initiative. Public protests against officials as well as "petty" corruption are raising the risks taken by those who are culpable.

Africa's links with the world are strengthening and expanding, and its growing trade with China is not only rehabilitating infrastructure that decayed during the Cold War period but augmenting it. The lessons of the export-oriented surface transport systems of the colonial era are not lost on Africa's modern planners, who negotiate with their Chinese partners to secure new and improved internal routes to stimulate local economies and facilitate market links.

Africa's millennia of misfortune have not diminished the vibrancy of its cultures nor its potential in a world to which it gave cultural rise. For tens of thousands of years Africa nourished and forged humanity, launching emigrations that would forever change the planet. Africa's time and turn will come again.

EPILOGUE: MORE THAN EVER

WHAT ARE THE LESSONS OF GLOBAL GEOGRAPHY THAT HELP US NAVigate the waves of change that sweep across the world? Here's one: our planet is not only small, it is also relatively young and geologically mobile. Its rocky crust is thin and split by belts of volcanic and seismic activity. Human history is replete with the consequences: Toba, Thera, Krakatoa, Tangshan. Between them, the twin earthquakes of Tangshan in 1976 and the tsunamis generated by the submarine quakes off Sumatra in 2004 killed about one million people. You would think that the planners and builders of nuclear power plants in the Pacific's "Ring of Fire" (see page 103) would heed such warnings of nature. Yes, there would have been substantial casualties from Japan's March 2011 Tohoku earthquake in any case. But the widespread radioactive contamination from nuclear reactors damaged and inundated by the resulting tsunamis could have been greatly reduced by better planning based on the physical and human geography of vulnerable Japan.

And here's another: distant, even remote places can have a powerful impact on anyone, anywhere. In America it took some time for the reality to sink in that the terrorist attacks of 9/11 had been planned in Taliban-ruled Afghanistan. In Russia in 1991 the name Chechnya was just another frontier "republic" somewhere in Caucasia. A few years later Chechen terrorists were sowing fear and mayhem in Moscow and beyond. Fishing boats capsized off the shores of East Africa, drowning their crews in waves caused by the Sumatra tsunami originating across the Indian Ocean thousands of kilometers away. And in late 2011 ocean-

---

TAKING GEOGRAPHY SERIOUSLY . . .

"Make no mistake," a spokesman for the Bangladesh secondary school examination board told shocked reporters at a press conference in Dhaka, "we are determined to stamp out cheating during examinations, and these isolated incidents will only strengthen our resolve."

During questioning the spokesman admitted that thousands of students had to be expelled for cheating or for attacking monitors and that dozens of teachers were suspended for illegally helping their students. In one college students set off home-made bombs to prevent an examination from taking place and "savage fighting" was reported to have occurred in a number of schools.

Asked who was responsible for such behavior, he replied: "I blame the parents, literally, because so many parents insist in sitting with their children during exams and helping them with the answers . . . the ones sitting the geography papers are the worst. They're vicious."

—*Bangladesh Daily News*, quoted in *Interaction*, 1992, 23:4.

---

ographers were tracking a mass of millions of tons of debris from earthquake-struck Japan on its way to the beaches of Hawaii and the coastline of the U.S. Pacific Northwest. To paraphrase an old saying about Las Vegas, what happens in Japan does not stay in Japan. Or Afghanistan. Or Chechnya.

Inevitably, events of this kind suddenly throw barely-known locales into high relief, jarring our consciousness and often making us wonder why we hadn't known about, or anticipated, what is happening. To be sure: geographers may be more map-conscious than other scholars, and there always are specialists among us who happen to have worked in such places; but geographers, too, have their reality checks. When the former Yugoslavia collapsed and such names as Krajina, Slavonia and Vojvodina made the news, many of us reached for our atlases. But what geography can contribute to the discourse is context. When the American intervention in Sunni-minority-ruled Iraq began, the prospect that this would lead to a potentially ungovernable but Shi'ite-dominated state, after the United States had supported Iraq in its war of the 1980s against Shi'ite Iran, appeared to constitute a massive geographic-strategic contradiction. Locations such as the Sunni Triangle and the Green Zone enhanced our mental map, but the bigger picture had to do with the potentially largest winner in this campaign: terrorist-supporting Iran. It still looks that way in early 2012, pending resolution of the evolving situation in Shi'ite-minority-ruled Syria.

Which brings us to another of geography's instructive pointers. No matter how transnational the world's challenges, nor how pervasive the forces of globalization, nothing has happened over the past half century to diminish the role and primacy, in international affairs, of the state. Iran may have shed its Persian past, but not its imperial imperatives, now expressed through its support for terrorist initiatives, its apparent determination to create nuclear capability for peace and war, and its implacable enmity toward Israel. To confront these issues, all parties have to deal with the Iranian machinery of state. In the international arena, Iran has its allies, and the most effective of these in recent years have been Russia and China. Iran may not be a large state territorially or demographically, but its influence outstrips its dimensions.

Political geographers after World War II liked to measure a state's power in terms of its capacity to use its resources to affect the behavior of other states, and there was a time when size mattered. Today a country smaller than Alabama with a population the size of Texas can threaten neighbors near and far through nuclear weaponry and missile technology, as North Korea proves. Rogue states, malfunctioning states, failed states—they are all part of the global political mosaic and they are probably safer from intervention today than they ever were—even as they pose a growing threat.

There was a time, not long ago, when it seemed as if the stable, representatively governed, internationally connected state, the building block of the international community, was in decline. The unification effort in Europe and the collapse of the U.S.S.R and Yugoslavia gave rise to notions of a New World Order in which the state would lose its dominance to unions, commonwealths, associations, "core" countries with their own peripheral spheres of influence, and other hypothetical constructs. It didn't happen: to the contrary, the number of sovereign states in the global political-geographical mosaic continued to grow. United Nations membership now approaches 200, nearly *four times* as many members as signed the founding treaty in 1945. None of these states is going to give up its membership after joining some multinational union. Notice that each of the 28 member-states of the EU has its own seat at the UN. None volunteered to yield it upon joining Europe's in-group.

To be geographically literate, do you have to know something diagnostic about every one of these 200 states? No, but to avoid a "Kissinger moment" of the kind quoted on page 18 it is useful to look at the regional layout on pp. 156–157 and to devise a mental map of states that matter. Proportionately, a far larger number of states today are peripheral, in every sense of that term, than was the case after the end of the Second World War. Timor-Leste, Swaziland, Gambia, Dominica, Tuvalu, and several dozen other mini- or microstates, many of them resulting from conspiracies of history and geography, should not clutter one's mental map. But what about the likes of Mauritius and Mauritania, the subjects of that embarrassing Oval-Office gaffe cited in Chapter 1? Well, by virtue of its proximity to Africa, the island state of Mauritius is often cited as that realm's highest-income country without oil, and Mauritania lies in a contentious zone near the western end of Africa's Islamic Front. That said, they don't exactly rank with the world's movers and shakers.

One way to let the map be the guide is to give it some time to sink in. I often ask my students to take a full-page map and spend as much time "reading" it as they would take with a page of print. That can lead to some interesting questions. Look at Brazil-dominated South America: only two countries are *not* neighbors of this emerging regional superpower, and if Argentina looks like the continent's second-place state, well, in many ways it is. In Middle America, Mexico is far larger than all the other countries and islands of the realm combined, a formidable neighbor whose troubles are spilling over into the United States, and vice versa (to quote a former Mexican president: "Poor Mexico. So far from God and so close to the United States").

Certainly South Asia merits a deep mental imprint. Not only is India on its way to becoming the world's most populous state: it already is the world's largest democracy. But *what* a neighborhood Hindu-majority India is in! Nuclear-armed, Islamic Pakistan to the west; teeming, also Muslim-dominated Bangladesh to the east with a combined population larger than the United States. Malfunctioning Nepal to the north and China across a contentious border; civil-war-plagued Sri Lanka to the south. But here's what really matters in India's geography: its modest territorial size. With a population soon to match China's, it is only one-third as large as its Asian neighbor and about the size of the United States east of the Mississippi. Think about it: 1.3 billion people between the Mississippi and the Atlantic! In how many different ways does this spell vulnerability?

One of the most interesting maps is the one of Southeast Asia, where Buddhist Thailand was once the leading state but Muslim Indonesia—spread across more than 17,000 islands—is taking its place as the regional leader. And take a look at what happens on that long Malay Peninsula stretching from near Bangkok all the way to Singapore. Four countries vie for space—and watch the news on the least-familiar one, Myanmar, formerly called Burma. In early 2012 it looked as though a new era could be dawning there after decades of disastrous military dictatorship.

On the other flank of Asia, Turkey's time has also come, a fast-growing, diversified economy underpinning regional ambitions dormant since Ottoman times. Here is another state to remind us always to look at the neighborhood: Turkey lies at the juncture of the Arab world, Transcaucasia, and Europe, with Kurdish problems to the south, Armenian ones to the east, and Greek ones to the west. But Sunni Islam is Turkey's majority religion, creating an opportunity for the Turks to set political as well as economic examples for the Muslim world to follow.

If Central Asia looks like a jumble on the map, it isn't any worse than Europe. No Germany or France has emerged here, but in this second decade there are signs that Russian interest in this region is reviving, in part as a counterweight to the difficulties Moscow has been having in Transcaucasia and the cold shoulder it gets from Europe. If domestic problems do not preoccupy the Putin regime set to take control in mid-2012, get ready for a massive Russian effort to reconstitute that failed post-Soviet Commonwealth of Independent States (CIS)—except that the vision this time will be grander. From Kazakhstan to Belarus and from Kyrgyzstan to Ukraine (yes, Ukraine), Putin is reported to envisage a counterweight to the malfunctioning European Union. In Moscow this will revive unhappy memories of the Soviets' failed intervention in Afghanistan, but Russia is in a far better economic position than the Soviet Union was in the 1980s, and Russian influence in the former Central Asian republics is on the rise. Still, this is a rough neighborhood: how would you like to be wedged between Pakistan and Iran or Turkmenistan and Tajikistan, with India looming over your shoulder? Just ask the Afghans. And here's a not-so-trivial question: does Afghanistan have a common border with China?

It is obviously useful to strengthen one's mental map of the geopolitical world. And it doesn't present any greater challenge, surely, than geology students face when getting acquainted with the "calendar" on page 107 or chemistry students when confronted with the periodic table of the elements? American students, like the general public, find themselves bombarded with information about past presidents and chronologies of historic events—how about a longer and more global geographic look at the present and the future? When "national" economies are inextricably linked, shouldn't we have greater awareness of the fundamentals of the nations involved? Were those investors really unaware of a Greek propensity for early retirement, tax evasion, fiscal caprice, and corruption (Greece ranks among Europe's worst-afflicted countries) when they bet on Athens as a Euro Zone member and then concluded that bailouts would rescue a society notoriously set in its welfare ways?

Certainly the adage that old habits die hard applies to states, and Greece is only one example. It underscores the fact that efforts—even massive, coordinated and international campaigns—to intervene and transform national cultures tend to be doomed. In the aftermath of Haiti's devastating 2010 earthquake that killed and injured more than a half million of the country's 10 million people, the international community poured billions of dollars of aid into a country that, in UN Special Envoy Bill Clinton's words, should "build back better." But the fresh start that tens of thousands of relief workers hoped for never materialized. A year after the quake, only about 15 percent of the available funds had been used for rebuilding, less than 5 percent of the rubble had been cleared, more than 1 million Haitians continued to live in makeshift shelters and tents even when their dwellings were ready for occupancy, and chaos and corruption afflicted the entire relief effort. Relief workers complained of weak, seemingly irremediable institutions, a lack of will and discipline, and a general failure of Haitians to pull together to ensure the progress that seemed to lie within reach.

If a massive international effort cannot even begin to achieve its goals in a small country such as Haiti, what are the chances in Iraq, or in Afghanistan? Cultural and physical geography constitute formidable adversaries even when the dollar figures are written in trillions and the human investment in the hundreds of thousands. It is often said these days that the failure of the United States to satisfactorily reconstitute Iraq and adequately pacify Afghanistan signals a weakening of American superpower capacities, but that is nonsense. Military surges can temporarily improve the tactical situation, but it would take millions of troops and other personnel, and quadrillions of dollars, to effectively control Iraq's historic centrifugal forces while saving Afghanistan from itself. Those who compare America's rebuilding successes after World War II in Japan and West Germany to the campaigns in Iraq and Afghanistan miss a key geographic difference: West Germany and Japan were not fractured by ethnic or religious differences and conflicts, and American and allied forces and other personnel did not have to keep warring factions apart while negotiating with terrorists in order to nation-build in hostile environments. What Iraq and Afghanistan, and, in different contexts, Haiti and Rwanda prove is that the era of superpower hegemony is over. Even in a failing state, as the British, the Soviets and the Americans have learned, habits, customs and mores of old cannot be suppressed and the cost of trying it is unaffordable. Change, when it happens, must come from within, as it did in South Korea, in China, and in South Africa—and now, perhaps, in some countries of the Arab World and in Myanmar.

Undoubtedly the United States will involve itself in progressive movements of the kind that are transforming several Arab states, but the Libyan model is more likely to prevail than the Iraqi one. The (second) Bush administration's obsession with implanting democracy in unfamiliar and unready settings has been replaced by less aggressive support for more representative government in places where the opportunity arises, although it remains to be seen whether regionally and culturally disparate Libya can achieve stability and build the institutions needed to support a form of popularly endorsed administration in the wake of its 2011 revolution. When Libya erupted even as the United States was still engaged in Iraq and Afghanistan, questions arose about this country's ability to prosecute a third war. A chorus of critics found an echo in the Obama administration: further nation-building should concentrate on the home front. And so the United States played a crucial but low-profile role in NATO's air campaign against Qaddafi's clique.

A decade ago, when we were contemplating the post-9/11 world in terms of the steadying, security-ensuring role of the United States, the prospect of superpower competition arising from China's ascent on the global stage, and the rise of "regional" powers such Brazil and India, a New World Order did seem to be in the offing. A contemplative epilogue in an earlier book concluded that "In the aftermath of World War II, the United States was the indispensable global presence. The transformation of China and the reconstruction of Russia . . . were enabled by America's stabilizing influence. The capacity of the United States to meet the challenges now looming will shape the future of this dangerous world" (de

Blij, 2005). Less than a decade later, superpower impotence is a more common refrain, and not only because the United States saw its capacities limited by logistical overreach and economic mismanagement. For all its rapid growth, huge reserves, and global reach, China too faced limitations financial and managerial—the former displayed in Europe in November 2011 when Beijing had to forgo an opportunity to buy influence in Europe, and the latter in evidence in Tibet and Xinjiang where China's repressive rule evoked responses similar to those arising in other parts of South and Southwest Asia invaded by foreigners.

If the political geography of this fast-changing world presents challenges to anyone trying to keep up with it, environmental issues are no less daunting even though, in the past several years, they have ranked lower among national priorities in America. In the 1990s, when scientists sounded the first alarms about global warming trends, there were reasons to withhold judgment on the durability of the track, for reasons cogently summarized by Jared Diamond in his book *Collapse* (Diamond, 2005). By the time the evidence had become incontrovertible, an ugly debate had politicized the science and confused the public. And when the scientific consensus should have been persuasive, the economic downturn that began in 2008 put other, more immediate priorities ahead of climate change.

Did weak geographic literacy have something to do with this? Reviewing the sequence of events it is not difficult to hear an echo of Robert McNamara's complaint that the American public was not sufficiently informed about the reasons and objectives behind the U.S. intervention in Vietnam. By the time the global-warming debate erupted, geographic education had long been at a low ebb, and, truth be told, scientists themselves did not do an adequate job of transmitting to lay audiences the basics needed to understand the complex issues involved. In the process, a huge opportunity to secure public interest and support for mitigation initiatives was lost: the link between climate change and public health. As noted earlier, nothing the planet's human occupants can do will "stop" climate change. Nevertheless, the gigantic tonnage of pollutants poured into the atmosphere every day creates present and future health hazards that must be countered. International action on this front would have climatic as well as health results, both favorable. But by the time this link might have enhanced public perception and support, the global recession consigned the associated expenditures to the status of unaffordable luxury.

As is becoming clearer every day, climate change of the kind the planet is experiencing today entails abrupt environmental swings that can have severe social and political consequences with consequences for national security and even global stability. It has the potential to disrupt order, threaten cities, dislocate millions, alter trade routes, affect water and food supplies. Studies of this aspect of the climate-change issue urge the creation of an international crisis-response mechanism that would coordinate the responsibilities of all participants' military and security operations, for activation in the event of an environmental emergency of the magnitude of, say, a Tambora (or Toba) event or the kind of sudden glacial surge that happened with the onset of the Younger Dryas.

National as well as personal security have come to the forefront of people's concerns, and not only because of the implications of the 9/11 terrorist attack. Over several decades I have kept informal notes on the conversations I got into with travelers on airplanes and in other settings, especially during the decade when I was on network television and often did not need to explain what I do for a living. In the 1960s and 70s, telling a fellow traveler that I am a geographer would most often lead to a trivia test: "Oh yeah? Well, where's Ouagadougou? What's the capital of Bhutan?" In the public image a geographer, it seemed, must be a walking gazetteer, a fount of facts. Then, during the late 1970s and early 80s, perhaps as a result of the energy crisis during the Carter administration, questions turned to commodities. Where could we get more oil? Why would the United States sell grain to Muslim countries that rationed their oil exports to America? But even before 2001, the focus changed again, to matters of security. Perhaps the random airplane hijackings and terrorist incidents that were on the rise were a factor, but it was noteworthy that people would ask where it would be safest to live on this planet, not just away from violence but also relatively safe from environmental threats such as earthquakes, tornadoes and hurricanes. "Is there a map of global risk?" was a question that came up in various contexts from time to time.

This seemed encouraging to me—my fellow travelers used to think of geography as almanac, next as inventory, now as analysis, with human as well as natural factors in the matrix. After 9/11 personal security issues took center stage, but more recently the economic-security implications of China's global ascent have come to the fore. One reward of taking three to four dozen trips a year is that you meet many people who have personal and professional experience in other countries and regions of the world, and who share a concern about the asymmetry of geographic knowledge between "us and them." As one of my seatmates said a few months ago, "every Chinese high-school graduate seems to be an accomplished geopolitician with a strong sense of nationalism. I get an earful every time I'm over there."

Given the inevitable uncertainties arising in a world in economic, cultural, political and environmental transition, it is not surprising that security is a primary concern among those tracking the big picture. China's global campaign to purchase, lease, or otherwise acquire millions of hectares of farmland in diverse environments signals uncertainty over its ability to provide food to 1.4 billion people in the event of abrupt climate change. India worries about snowfall patterns in the Himalayan Mountains that keep its lifeline rivers flowing. Americans argue over oil and gas exploitation and pipeline construction in ecologically sensitive environments at a time when distant supply lines are subject to sudden disruption. Iran couples its growing nuclear capacity with a stated intent to annihilate its adversary Israel, threatening to bring to an end 65 years of proliferation coupled with prudence. A regional rebellion is shaking the foundations of Arab society from Syria to Yemen and from Libya to Bahrain, with unpredictable outcomes for governmental, religious, and military interests. The European Union is at a crossroads, its grand design at risk from economic and political corners.

These interconnected circumstances give us a glimpse of the world ahead. Barring setbacks that can afflict any state growing economically as rapidly—and regionally as disparately—as China currently is, it may well be that China's economy will become the world's largest before 2030. But economic success does not guarantee that China's political system is durable or that its social norms—notably its Hanification practices—will have a place in a globalizing world. Lamentations to the contrary notwithstanding, the United States will continue to be the cornerstone of the international state system, although its primacy is eroding. America shares common ground with many of the world's states in arenas ranging from representative government to human rights and from multicultural nationhood to open society. In the world ahead, U.S. leadership will remain indispensable; its natural and historic allies, from India to Brazil and from Europe to Indonesia, will become increasingly important on the world stage. This presents opportunities to shape a new world order, a global initiative to center on the strengthening of voluntary relationships between the United States and international partners that share its vision of the future. Such an enterprise should take precedence over all others, a grand geographic design built on foundations tried and tested, and upon which the world depends—now more than ever.

WORKS CITED

9/11 Commission, 2004. *Final Report of the National Commission on Terrorist Attacks Upon the United States.* New York: Norton.

Alley, R. B., 2002. *Abrupt Climate Change.* Washington, D.C.: National Academy Press.

Altman, L. K., 2002. "By 2010, AIDS May Leave 20 Million Orphans." *New York Times,* July 10.

Annan, K., 2004. "Courage to Fulfill Our Responsibilities." *Economist,* 12/4: 21.

Ardrey, R., 1966. *The Territorial Imperative.* New York: Atheneum Press.

Aryeetey-Attoh, S., Ed., 1997. *Geography of Subsaharan Africa.* Upper Saddle River: Prentice-Hall.

Baker, J. H., 1996. *Russia and the Post-Soviet Scene: A Geographical Perspective.* Hoboken: John Wiley & Sons.

Banchero, S., 2011. "Don't Know Much About Geography." *Wall Street Journal,* July 20.

Begun, D. R., 2003. "Planet of the Apes." *Scientific American,* 289: 2.

Berry, B. J. L., 1999. "Deja vu, Mr. Krugman." *Urban Geography* 20/ I: 1.

Best, A. C. G., and H. J. de Blij, 1977. *African Survey.* New York: Wiley.

Botkin, D., 2009. "Taking Steps to Fight Global Warming." Letter, *New York Times,* July 3.

Bremer, P. L., 2011. "Iraq's Tenuous Post-American Future." *Wall Street Journal,* December 28.

Carmody, P., 2011. *The New Scramble for Africa.* Cambridge: Polity.

Charlemagne, 2003. "Europe's Population Implosion." *Economist,* 7/19: 42.

Clarke, R. A., 2004. *Against All Enemies.* New York: Free Press, 40.

Cohen, J. E., 2003. "Human Population: The Next Half Century." *Science,* 302: 14, 1172.

Collier, P., 2007. The Bottom Billion: Why the Poorest Countries Are Failing and What Can Be Done About It. New York: Oxford University Press.

Curtin, P., 1969. *The Atlantic Slave Trade.* Madison: University of Wisconsin Press.

Cutter, S. L., D. B. Richardson, T. J. Wilbanks, Eds., 2003. *The Geographic Dimensions of Terrorism.* New York: Routledge.

Davis, C. S., 2004. *Middle East for Dummies.* Hoboken: John Wiley & Sons.

Davis, S., 2002. *The Russian Far East: The Last Frontier.* New York: Routledge.

de Blij, H. J., 1971. *Geography: Regions and Concepts.* New York: Wiley.

de Blij, H. J., 1991. "Africa's Geomosaic Under Stress." *Journal of Geography*, 90: 1.

de Blij, H. J., 1996. "Tracking the Maps of Aggression." *New York Times*, October 28.

de Blij, H. J., 2003. "Seeking Common Ground on Iraq." *New York Times*, October 11.

de Blij, H. J., and P. O. Muller, 2006. *Geography: Realms, Regions and Concepts*, 12th ed. Hoboken: John Wiley & Sons.

de Blij, H. J., 2004. "Africa's Unequaled Geographic Misfortunes." *Pennsylvania Geographer* 42/1: 3–28.

de Blij, H. J., Ed., 2005. *Atlas of North America.* New York: Oxford University Press.

de Blij, H. J., P. O. Muller, J. Nijman, 2012. *Geography: Realms, Regions, and Concepts.* Hoboken: John Wiley & Sons.

Demko, G. J. et al., Eds., 1998. *Population Under Duress: The Geodemography of Post-Soviet Russia.* Boulder: Westview Press.

Deutsch, D., 2011. *The Beginning of Infinity.* New York: Viking.

Diamond, J., 1997. *Guns, Germs, and Steel.* New York: Norton.

Diamond, J., 2005. *Collapse: How Societies Choose to Fail or Succeed.* New York: Viking.

Economist, 2003, "Can Oil Ever Help the Poor?" 12/4: 42.

Economist, 2004a, "The Ultra-Liberal Socialist Constitution," 9/18: 59.

Economist, 2004b, "Why Europe Must Say Yes to Turkey," 9/18: 14.

Economist, 2004c, "The Impossibility of Saying No," 9/18: 30.

Eltis, D., and Richardson, D., 2010. *Atlas of the Transatlantic Slave Trade.* New Haven: Yale University Press.

Fagan, B., 2000. *The Little Ice Age.* New York: Basic Books.

Friedman, T., 1996. "Your Mission, Should You Accept It." *New York Times*, October 27.

Friedman, T., 2005. *The World Is Flat.* New York: Farrar, Straus & Giroux.

Gewin, V., 2004. "Mapping Opportunities." *Nature*, 247: 376.

Gould, P. R., 1993. *The Slow Plague: A Geography of the AIDS Pandemic.* Cambridge: Blackwell.

Graham, R., and J. Nussbaum, 2004. *Intelligence Matters.* New York: Random House.

Grosvenor, G. M, 1988. "Why Americans Don't Know About Geography and Why It's Hurting Us." Washington, D.C.: National Press Club.

Grosvenor, G. M., 1995. "Geography in the Public Eye." *Washington Post,* July 28.

Grove, J. M., 2004. *The Little Ice Age*, 2nd ed. London: Methuen.

Hall, S. S., 1993. *Mapping the Next Millennium.* New York: Random House.

Hansen, J., 2004. "Defusing the Global Warming Time Bomb." *Scientific American*, March, 68–77.

Harris, S., 2004. *The End of Faith: Religion, Terror, and the Future of Reason.* New York: Norton.

Harvey, D., 2004. *The New Imperialism*. Oxford: Oxford University Press.

Harvey, R., 2003. *Global Disorder: America and the Threat of World Conflict*. New York: Carroll & Graf.

Hertslet, E., 1909. *The Map of Africa by Treaty*. London: His Majesty's Stationery Office.

Hitchcock, W. I., 2002. *The Struggle for Europe*. New York: Doubleday.

Horgan, J., 1996. *The End of Science: Facing the Limits of Knowledge in the Twilight of the Scientific Age*. Reading: Addison-Wesley.

Hu, W., 2011. "Geography Report Card Finds Students Lagging." *New York Times*, July 20.

Huntington, E., 1942. *Principles of Human Geography*. New York: Wiley.

Huntington, E., 1945. *Mainsprings of Civilization*. New Haven: Yale University Press.

Huntington, S. P., 1996. *The Clash of Civilizations and the Remaking of the World Order*. New York: Simon & Schuster.

Jacques, M., 2009. *When China Rules the World*. New York: Penguin.

Kagan, R., 2004. *Of Paradise and Power*. New York: Vintage.

Kaiser, R. J., 1994. *The Geography of Nationalism in Russia and the U.S.S.R*. Princeton: Princeton University Press.

Kaplan, R., 2000. *The Coming Anarchy*. New York: Random House.

Kasting, J. F., 2004. "When Methane Made Climate." *Scientific American*, 291/1: 78.

Kepel, G., 2002. *Jihad: The Trial of Political Islam*. Cambridge: Harvard University Press/Belknap.

Kepel, G., 2004. *The War for Muslim Minds and the West*. Cambridge: Harvard University Press/Belknap.

Kerr, R. A., 2003. "Has an Impact Done it Again?" *Science*, 302: 21.

Kinzer, S., 1996. *Overthrow: America's Century of Regime Change from Hawaii to Iraq*. New York: Times Books.

Krugman, P., 2009. "Betraying the Planet." Column, *New York Times*, June 29.

Langewiesche, W., 2004. *The Outlaw Sea*. New York: North Point Press.

Lincoln, B., 1994. *The Conquest of a Continent: Siberia and the Russians*. New York: Random House.

Mackinder, H. J., 1904. "The Geographical Pivot of History." *Geographical Journal*, 23: 421–37.

Martin, G., 2005. *All Possible Worlds: A History of Geographical Ideas*. New York: Oxford University Press.

McCune, S., 1956. *Korea's Heritage: A Regional and Social Geography*. Rutledge, VT: Charles Tuttle.

McNamara, R. S., 1995. *In Retrospect: The Tragedy and Lessons of Vietnam*. New York: Times Books.

Monmonier, M., 1991. *How to Lie With Maps*. Chicago: University of Chicago Press.

Monmonier, M., 1997. *Cartographies of Danger*. Chicago: University of Chicago Press.

Moss, Walter, 2008. *An Age of Progress? Clashing Twentieth-Century Global Forces.* London: Anthem Press.

Muehrcke, P. C., and J. O. Muehrcke, 1997. *Map Use: Reading-Analysis-Interpretation,* 4th ed. Madison: JP Publishers.

Mueller, J. E., 2009. *Overblown: How Politicians and the Terrorism Industry Inflate National Security Threats, and Why We Believe Them.* New York: Free Press.

Murck, B. W. and B. J. Skinner, 1999. *Geology Today.* New York: John Wiley & Sons.

Murdock, G. P., 1959. *Africa: Its Peoples and Their Culture History.* New York: McGraw-Hill.

Myers, S. L., 2004. "Putin Gambles on Raw Power." *New York Times,* September 19.

National Geographic, 2005. *Atlas of the World,* 8th ed. Washington, D.C.: National Geographic Society.

Newman, J., 1995. *The Peopling of Africa: A Geographical Interpretation.* New Haven: Yale University Press.

Onishi, N., 2001. "Rising Muslim Power in Africa Causes Unrest in Nigeria and Elsewhere." *New York Times,* November 1.

Oppenheimer, S., 2003. *The Real Eve: Modern Man's Journey Out of Africa.* New York: Carroll and Graf.

Oxford University Press, 2004. *Atlas of the World,* 12th ed. New York: Oxford University Press.

Patten, C., 1998. *East and West: The Last Governor of Hong Kong on Power, Freedom, and the Future.* London: Macmillan.

Qiang, S., 1999. *China Can Say No.* Beijing: National Press.

Rachman, G., 2004. "Outgrowing the Union." *Economist,* Survey of the European Union, 9/25: 3.

Reader, J., 1998. *Africa: Biography of a Continent.* New York: Knopf.

Reardon, S., 2011. "Geographers Had Predicted Osama's Possible Whereabouts," *Science,* May 2.

Remnick, D., 1993. *Lenin's Tomb: The Last Days of the Soviet Empire.* New York: Random House.

Rifkin, J., 2004. *The European Dream.* New York: Tarcher/Penguin.

Robinson, A. H., and B. Petchenik, 1976. *The Nature of Maps: Essays Toward Understanding Maps and Mapping.* Chicago: University of Chicago Press.

Rohter, L., 2002a. "Argentine Judge Indicts Four Iranian Officials in 1994 Bombing of Jewish Center." *New York Times,* March 10.

Rohter, L., 2002b. "South America Under Watch for Signs of Terrorists." *New York Times,* December 15.

Sachs, J. D., 2000. *The Geography of Economic Development.* Newport, RI: United States Naval War College Jerome E. Levy Occasional Paper in Economic Geography and World Order.

Schmitt, E., and T. Shanker, 2004. "Estimates by U.S. See More Rebels with More Funds." *New York Times,* October 22.

Shaw, D. J. B., Ed., 1999. *Russia in the Modern World: A New Geography.* Malden: Blackwell.

Solis, P., 2004. "AAG Member Profile: E. 'Fritz' Nelson." *AAG Newsletter,* 39/1: 9.

Spykman, N. J., 1944. *The Geography of the Peace.* New York: Harcourt, Brace & Co.

Stanley, W. R., 2001. "Russia's Kaliningrad: Report on the Transformation of a Former German Landscape." *Pennsylvania Geographer,* 39: 1.

Stock, R., 2004. *Africa South of the Sahara: A Geographical Interpretation,* 2nd ed., New York: Guilford Press.

Terrill, R., 2003. *The New Chinese Empire: What it Means for the United States.* New York: Basic Books.

Thornton, J., and C. E. Ziegler, Eds., 2002. *Russia's Far East: A Region at Risk.* Seattle: University of Washington Press.

Tishkov, V. 2004. *Chechnya: Life in a War Torn Society.* Berkeley: University of California Press.

Trenin, D., 2002. *The End of Eurasia: Russia on the Border Between Geopolitics and Globalization.* Washington: Carnegie Endowment for International Peace.

Van Natta Jr., D., and L. Bergman, 2005. "Militant Imams Under Scrutiny Across Europe." *New York Times,* October 25.

Veregin, H., Ed., 2005. *Goode's World Atlas.* Skokie: RandMcNally.

Vogel, E., 2011. *Deng Xiaoping and the Transformation of China.* Cambridge: Harvard University Press.

Wegener, A., 1915. *The Origins of Continents and Oceans.* New York: Dover (reprint of the 1915 original, trans. by John Biram, 1966).

Weinberger, C. W., 1989. "Bring Back Geography." *Forbes Magazine,* December 25.

Wilford, J. H., 2001. "A Science Writer's View of Geography." New York: Opening Session, Association of American Geographers.

Wilford, J. N., 1981. *The Mapmakers.* New York: Knopf.

Wilson, E. O., 1995. *Naturalist.* Washington, D.C.: Island Press.

Wright, R., 1986. "Islamic Jihad." *Encyclopaedia Britannica Book of the Year.* Edinburgh: Encyclopaedia Britannica, Inc.

INDEX

Taiwan and, 211, 215–17
terrorism in, 160
in Vietnam War, vii, 16, 18, 29, 59,
69, 150–54, 182–83, 208
West Against the Rest and, 158
United States Geological Survey
(USGS), 41, 42*f*
University of California–Berkeley, 11
University of Miami, 20–22
Ural Mountains, 274
urbanization, 71, 81, 84, 168, 220
U.S. *See* United States
USA Today, 48, 51
USGS. *See* United States Geological
Survey
U.S.S. Cole, 203
USSR. *See* Soviet Union
Usuli doctrines, 178
Uzbeks, 185–86

Vakhan Corridor, 172, 173*f*, 188
Van Loon, Hendrik Willem, 5–6
Van Rompuy, Herman, 259
Venezuela, 207
Vietnam War
Afghanistan compared to, 182–83
U.S. in, vii, 16, 18, 29, 59, 69, 150–54,
182–83, 208
volcanoes
Kamchatka Peninsula, 275
Laki, 132
Ring of Fire, 103, 103*f*, 320
Tambora, 133–34
Thira, 128
Toba, 120–21

al-Wahhab, Muhammad ibn Abd, 163
Wahhabism, 162–63, 167–68
war. *See also* Cold War
civilizational clashes, 155, 157–58
fault line, 157
First Gulf, 154–55, 164, 169, 178
First Opium, 222
geographic realms and, 155–58,
156*f*–157*f*
geography and, 18, 148–88,
156*f*–157*f*
Indochina, 150–54, 208
Iran-Iraq, 169, 179

Iraq, 18, 29–31, 69, 174–81, 183–84,
239, 321
Islam and, 158
Korean, 149–50, 211
1967, 192–93
technology, 39
Vietnam, vii, 16, 18, 29, 59, 69,
150–54, 182–83, 208
World War II, 5, 19, 65, 148, 159–60,
214
Waziristan, 172, 182, 184, 203, 207
weapons of mass destruction (WMD),
29, 207
weather, on maps, 138*f*, 140–47,
142*f*–143*f*
Wegener, Alfred, 62–64, 63*f*, 102–4,
102*f*
Weinberger, Caspar, 17
West Africa, 198, 201, 299–301. *See also*
specific countries
West Against the Rest, 157–58
Western Europe, 214, 242. *See also*
specific countries
Wilford, John N., 9, 28
Will, George, 138
Williams, G. Mennen, 59
Wilson, Edward O., 294
wine, geography of, 14
Wisconsinan Glaciation, 118–20,
122–23, 134–35, 136*f*
WMD. *See* weapons of mass
destruction
world atlas, ix, 33–34, 49, 52, 85,
247–48
world cities, 88
World Trade Center
first attack on, 164, 195
9/11 attack on, 171–72, 175, 189–90,
206
World War II, 5, 19, 65, 148, 159–60,
214

Xian (formerly Ch'angan), 128–29, 221
Xianggang (Hong Kong), 230
Xinjiang, China, 224, 231
Xizang (Tibet), 66, 216, 224, 233

Yalu River, 149
Yanukovych, Victor, 292